ACCOUNTING AND FINANCIAL MANAGEMENT FOR CONSTRUCTION

CONSTRUCTION MANAGEMENT AND ENGINEERING
Edited by John F. Peel Brahtz

ACCOUNTING AND FINANCIAL MANAGEMENT FOR CONSTRUCTION

CHARLES H. MOTT, PH.D., CPA
TOWSON STATE UNIVERSITY

A Wiley-Interscience Publication
JOHN WILEY & SONS
New York   Chichester   Brisbane   Toronto   Singapore

Library of Congress Cataloging in Publication Data:

Mott, Charles H., 1930-
    Accounting and financial management for con-
struction.

    (Construction management and engineering)
    "A Wiley-Interscience publication."
    Includes index.
    1. Construction industry—Accounting. 2. Con-
struction industry—Finance. I. Title. II. Series.

HF5686.B7M68   658.1'5   81-10502
ISBN 0-471-07959-6   AACR2

SERIES PREFACE

Industry observers agree that most construction practitioners do not fully exploit the state of the art. We concur in this general observation. Further, we have acted by directing this series of works on Construction Management and Engineering to the continuing education and reference needs of today's practitioners.

Our design is inspired by the burgeoning technologies of systems engineering, modern management, information systems, and industrial engineering. We believe that the latest developments in these areas will serve to close the state of the art gap if they are astutely considered by management and knowledgeably applied in operations with personnel, equipment, and materials.

When considering the pressures and constraints of the world economic environment, we recognize an increasing trend toward large-scale operations and greater complexity in the construction product. To improve productivity and maintain acceptable performance standards, today's construction practitioner must broaden his concept of innovation and seek to achieve excellence through knowledgeable utilization of the resources. Therefore our focus is on skills and disciplines that support productivity, quality, and optimization in all aspects of the total facility acquisition process and at all levels of the management hierarchy.

We distinctly believe our perspective to be aligned with current trends and changes that portend the future of the construction industry. The books in this series should serve particularly well as textbooks at the graduate and senior undergraduate levels in a university construction curriculum or continuing education program.

JOHN F. PEEL BRAHTZ

La Jolla California
February 1977

PREFACE

Almost all of the books about construction accounting and financial management are written from the perspective of the large worldwide construction firm. Although this book overlaps with others in some areas of methodology, it is written for the small to medium-large construction firm. It includes all of the functions peculiar to that segment of the construction industry.

This book may be used in training or for reference. It can serve to teach construction accounting in a construction certificate program at a university. The text begins with basic accounting for construction and includes questions for testing the user. Some readers may need a knowledge of high school bookkeeping. Others, with an aptitude for business, may need no preliminary work.

After a review of basic accounting techniques the book proceeds to describe accounting controls and procedures. Accounting principles and policies are discussed in Chapter 3. Purchasing, accounts payable, cash disbursements, the financial statements, and the fiscal period are included in the next two chapters. Chapter 6 is concerned with the integrated functions of invoicing, accounts receivable, and cash receipts. Operational assets and taxes are emphasized in Chapter 7.

Chapter 8 describes personnel policies, project direct and indirect labor, and other indirect construction costs. Construction project (job) costs and budgeting and management reports are the concern of the next two chapters. In Chapter 11 management information systems (data processing) and bonding are examined. Financial analyses are related to the construction industry in Chapter 12, followed by a summary and conclusions.

I welcome suggestions for improving the content and organization of this book.

CHARLES H. MOTT

Baltimore, Maryland
September 1981

ACKNOWLEDGMENTS

A manuscript cannot be prepared without the patience of those close to the author. I thank my wife Madeline and my daughter Karen for their indulgence and encouragement.

I am indebted to my neighbor, Larry Seng, for typing the manuscript and correcting misspellings. For those that remain and any other errors in the text, I take responsibility.

Also, I thank Dr. John F. Peel Brahtz, Dr. Julien Wade, Ardy Hickman, Pat Atkinson, Jeanne Blair, Arnetta Jackson, Amanda Keach, and Barbara Weaver for their encouragement and offers of help.

CHARLES H. MOTT

CONTENTS

CHAPTER 12   FINANCIAL ANALYSES   191

CHAPTER 13   SUMMARY AND CONCLUSIONS   205

INDEX   209

ACCOUNTING AND FINANCIAL MANAGEMENT FOR CONSTRUCTION

CHAPTER 1

INTRODUCTION AND REVIEW OF BASIC ACCOUNTING TECHNIQUES

Accounting is the language of business.[1] Through accounting reports and information the management of the firm measures the success or failure of their actions and decisions. Accounting data are designed to present a reasonably accurate economic picture of business performance. By application of accounting principles and techniques, the accountant attempts to produce information that, if properly used, will tell the reader about the progress of the firm.

To measure the success of business firms and to compare them the accountant uses the concept of profit or income. Although the purpose of a business is to take in more cash than it pays out, the uncertainty of cash flow requires interim measures of performance. Profit or income is the accountant's surrogate for cash flow. In later chapters more is said about profit or income as a surrogate for cash flow, as imperfect as it may be.

Business firms in the construction industry as well as other industries, out of necessity, rely on accounting concepts as a major source of communication. Publicly held firms are required to use accounting information to report to their shareholders and government agencies. Smaller firms and privately held firms are often required to report in a similar manner to meet the requirements of

---

[1]George J. Chorba, *Accounting For Managers* (New York: American Management Association Extension Institute, 1978), p. 1.

debt agreements or of loan applications. In the construction industry the latter circumstances are often the case.

Despite the increase of rules and regulations for the use and reporting of accounting information to outsiders, uniformity of use and reporting of accounting information is the exception rather than the rule. However, accounting information must, in addition to serving external needs, meet the needs of the internal management of the firm. The managers of a business firm need accounting information to evaluate their performance. Therefore, the accounting information system serves internal and external informational needs.

All of these issues are examined in this book as they relate to the construction industry. The examination begins with a review of the basic accounting formula and techniques.

THE BASIC ACCOUNTING FORMULA

The accounting formula is based on a double entry system. The double entry system means that there is a debit and a credit side to each accounting transaction. The purpose of the double entry system is to prevent mistakes and to record both the asset effect and the liability or capital effect of each entry. The double entry method does not prevent errors, but it prevents the recording of only one side of an entry. If both sides are not recorded the books of account do not balance and a search is required to locate the error.

The accounting formula can be expressed in at least two ways: (*a*) the total dollar value of assets equals the total dollar value of liabilities plus the total dollar value of the owner's equity accounts (capital accounts), assets = liabilities + capital); or (*b*) the total dollar value of assets less the total dollar value of liabilities equals the owner's equity accounts (capital accounts), (assets − liabilities = capital). An example of the use of the accounting formula is given in Exhibit 1.

ASSETS

Assets are part of the accounting formula and are equal to the total of the liabilities and capital accounts. The assets of a business firm are such things as the cash it has in the bank, its accounts receivable for customers, its buildings and equipment, and its inventory. The assets can be broadly defined as items or rights owned by the firm

that can be quantified. From this definition it can be seen that a firm's assets could consist of intangible as well as tangible items. A tangible asset could be a building or cash. An example of an intangible asset is a patent right or goodwill.

LIABILITIES

Liabilities are debts or amounts owed by the business entity. Such items as amounts owed to suppliers, to banks, and to employees are liabilities. Liabilities are liquidated by the use of assets, such as cash, the sale of marketable securities, or inventory. Examples of liabilities that would appear on a business firm's financial statements are accounts payable, notes payable, bonds payable, and mortgages and loans payable.

CAPITAL

The capital accounts are described by various names depending on the type of ownership of the firm. If the firm is owned by a single individual the capital account is described as proprietorship equity. If the firm is a partnership the capital accounts are described as partnership capital and drawing accounts. Firms that are owned by stockholders (shareholders) have capital accounts that are usually described as owner's equity. In a typical corporation these accounts are classified into three areas: (a) capital stock, (b) paid in capital in excess of par or stated value, and (c) retained earnings. Regardless of the descriptions and the number of accounts, the capital interest in a firm is a residual interest. This point was illustrated by the basic accounting formula described above. The assets less the liabilities is the residual that belongs to the firm's owners. Balance sheets (statements of financial position) of large companies, including large construction companies, have the following accounts in their owner's equity sections: (a) capital stock—common—class A (authorized 1,000,000 shares; issued and outstanding 500,000 shares), (b) capital stock—common—class B (authorized 500,000 shares; issued and outstanding 400,000 shares, (c) preferred stock—10%—cumulative (authorized 300,000 shares; issued and outstanding 200,000 shares), (d) paid in capital in excess of par—capital stock—common—class A, (e) paid in capital in excess of par—capital stock—common—class B, (f) retained earnings, (g) treasury stock (assuming that the cost method was used to account for the firm's purchase of its own stock).

THE ACCOUNTING DEFINITION OF PLUS AND MINUS

The accounting profession has developed its own techniques for recording increases and decreases in asset, liability, and equity accounts.[2] Asset accounts are increased by "debits" and decreased by "credits." Asset accounts usually have a "debit" balance. Liability and capital accounts are increased by "credits" and decreased by "debits" and usually have credit balances. This procedure is consistent with the accounting formula that assets = liabilities + capital.

Debits are defined as entries on the left side of journals, accounts, ledgers, and statements. Credits are defined as entries on the right side of journals, accounts, ledgers, and statements. These are comparable to plus and minus designations.

For asset accounts a debit is a plus and adds to the account balance, while a credit is a minus and subtracts from the account balance. Liability and capital accounts, because they are "right side" or credit accounts, are decreased by plus and increased by minus (credit) entries. Because expense and income accounts represent changes in owner's equity during an accounting period, debits increase expense accounts and decrease owner's equity. Credits, however, increase income and increase owner's equity. The objective is to have a net credit after subtracting expenses from the income accounts and thereby increase the owner's equity at the end of the accounting period. Remembering the relationship between asset, liability, and owner's equity accounts and debit and credit entries makes it easier to understand accounting "journal entries."

ACCOUNTING JOURNAL ENTRIES

The accounting journal is the book of original entry. Nothing is entered into the accounting ledgers and thus the financial statements unless it is recorded in a journal. The function of the accounting journal is to record any original entry into the accounting system. Journal entries consist of a date, usually an account number, an account description, the dollar amounts of the entry, and an explanation of the entry. Entries can be recorded in either a general journal or a specialized journal. If a general journal is the only journal used, all transactions are recorded in the general journal. If specialized journals are used in addition to a general

[2]Donald E. Kieso, and Jerry J. Weygandt, *Intermediate Accounting* (New York: John Wiley & Sons, 1977), p. 63.

journal, any entries that do not fit in the specialized journals are recorded in the general journal. An example of a journal entry is the recording of a cash payment received on account from a customer:

1979

		Debit	Credit
Jan. 31	Cash	$10,000	
	Accounts receivable		$10,000
	To record payment on account received from John Dow for the purchase of product x.		

INTERNAL CONTROL OF JOURNAL ENTRIES

Access to the general ledger must be controlled. Control is necessary to safeguard the firm's assets and to assure management that their financial statements are reasonably accurate. Because the journals are the books of original entry, protecting access to the ledger is accomplished through the control of journal entries. All journal entries to a general ledger are usually prenumbered. Journal entries which are prepared continuously, that is, in each accounting period, are given standard numbers which remain the same from period to period. Noncontinuous entries require a journal number before they can be accepted for entry into the general ledger. Therefore, the employee in charge of the general ledger knows which entries to expect and has control over access to the ledger to prevent compromise of the firm's assets.

This procedure gives the general ledger accountant control over accounting input to the ledger. In this way, the responsible employee can be sure that all entries to the general ledger have been recorded. Verifying the recording of journal entries increases the accuracy of the financial statements. An additional procedure is to require that all journal entries have the necessary approvals, which usually is the signature of the manager of the area where the journal entry originates or the signature of a representative of the chief accounting officer. Both procedures reduce the possibility of fraudulent journal entries to disguise poor stewardship, the misuse of the firm's assets, and the manipulation of the financial statements.

THE FLOW OF ACCOUNTING INFORMATION

The first step in the flow of accounting information is the occurrence of a recordable event. Remembering that accounting is the language of business, any event that affects the firm in monetary terms is a recordable event. The effect may be on the assets, liabilities, or the owner's equity, either directly or by affecting an income or expense account. The monetary effect may not necessarily be exact, that is, the amount may only be estimable. It is one of the requirements of accounting that events be recorded as long as the amounts involved can be estimated with reasonable accuracy. There may be other requirements to qualify an event as recordable or to determine when it is recordable. For example, to record a sale the accounting requirement of "exchange" (to be discussed later) must be met. Some examples of recordable events are: (a) the sale of merchandise to a customer, (b) the purchase of an asset such as a piece of equipment, a machine, or a building, (c) the purchase of raw material, parts, or supplies from a vendor (supplier), (d) the purchase of an insurance policy on assets or employees, (e) the payment or the liability to pay employees for services rendered, (f) the depreciation or the "using up" of the firm's equipment, machinery, and buildings, (g) the incurrence or payment of debt, (h) the sale of stock to owners, or (i) the declaration or payment of a dividend to shareholders.

JOURNAL ENTRIES FOR RECORDABLE EVENTS

A recordable event is put into the firm's accounting records through a journal entry. The journal entries for the recordable events included above are illustrated by the following examples. The order of entry of the information is: (a) journal entry date, (b) the account numbers, (c) the account descriptions, (d) the general ledger page reference of the account, and (e) the debit and credit amounts.

(a) Sep. 30	2001	Accounts receivable	GL3	$10,000	
	9501	Sales	GL10		$10,000
(b) Sep. 30	4001	Equipment	GL4	$20,000	
	6001	Accounts payable	GL7		$20,000
(c) Sep. 30	3001	Purchases or inventory	GL3	$ 5,000	
	6001	Accounts payable	GL7		$ 5,000
(d) Sep. 30	9601	Insurance expense	GL11	$ 1,000	
	6001	Accounts payable	GL7		$ 1,000
(e) Sep. 30	9701	Payroll expense	GL11	$90,000	
	6002	Payroll payable	GL7		$90,000
(f) Sep. 30	9801	Depreciation expense	GL12	$18,000	
	4101	Allowance for depreciation— equipment	GL4		$18,000

(g) Sep. 30 1001 Cash   GL1 $95,000
            7001   Bonds/payable   GL8 $95,000

(h) Sep. 30 2002 Accounts receivable   GL3 $80,000
            9001   Common stock-par   GL14 $50,000
            9002   Paid in capital in
                   excess of par—
                   common stock   GL14 $30,000

(i) Sep. 30 9103 Dividend expense   GL14 $11,000
            6003   Dividend payable   GL7 $11,000

In the journal entries illustrated, the date, account numbers, general ledger page numbers, and amounts were arbitrarily selected. However, the account numbers and general ledger page numbers were selected for consistency with the accounts used. The account and page numbers would result in a grouping of accounts by balance sheet (statement of financial position) classification. Omitted from the entries are the journal entry explanations which are an important part of the journal entry in actual practice.

THE GENERAL LEDGER

To summarize the accounting information directly from the journal entries would be too time consuming and a source of too many errors. Therefore, accountants use the general ledger as a method of accumulating entries to each account. The general ledger reduces the chance of error and enhances account analysis.

Within the general ledger each account is listed. The order of listing is assets, liabilities, owner's equity, expenses, and income accounts. Then the journal entries are posted, one at a time to the general ledger either as a debit or a credit to the appropriate general ledger account. In the particular journal an entry is made referencing the general ledger page to which the posting was made. After all journal entries are posted the general ledger is summarized by adding the debits and credits and calculating a balance for each account in the ledger. When each entry is posted in the general ledger a reference is made in the ledger account to the journal from which the posting was made. References are necessary in case of posting errors.

After the balances have been calculated for each account they are listed in general ledger order. The purpose of the listing is to make certain that the general ledger is in balance—that the total of the debit accounts equals the total of the credit accounts. This summarization of the general ledger is known as a trial balance.

After the first trial balance is completed and balanced, the closing and adjusting entries are prepared and posted. Then another trial balance is made. This last trial balance is called a post closing trial balance. (The preparation of trial balances is discussed in detail in later chapters.) From the post closing trial balance the financial statements are prepared. More is said about the financial statements in later chapters. However, four financial statements are usually prepared in accordance with generally acceptable accounting principles. The first statement is the statement of financial position, formerly the balance sheet; second, the income statement, formerly the profit and loss statement; third, the statement of retained earnings or the statement of owner's equity; and, fourth, the statement of changes in financial position, formerly the changes in working capital.

SUMMARY

This chapter discusses the role of accounting as the language of the business firm. The reader is introduced to the basic accounting formula and to some of the basic language of accounting. Accounting debits and credits are defined and related to the structure of the accounts. The use of and nature of accounting journal entries are described. The flow of accounting information is traced, from a recordable event through the journal entries. The posting to the general ledger is described as well as the summarizing of the general ledger, the preparation of the trial balance, preparation of the post closing trial balance, and the transcribing of the financial statements.

EXHIBIT 1.   Statement of Financial Position (Balance Sheet)

Assets

Current Assets

Cash	$	10,000.00
Marketable securities		60,000.00
Accounts receivable		100,000.00
Inventories		240,000.00
Total current assets	$	410,000.00

Tangible Operational Assets

Land	$	500,000.00
Plant and equipment		10,000,000.00
Allowance for depreciation		(2,000,000.00)

EXHIBIT 1. (Continued)

Total tangible operational assets	$ 8,500,000.00
Intangible Operational Assets	
Unexpired insurance	$ 5,000.00
Total assets	$ 8,915,000.00
Liabilities and Owner's Equity	
Current Liabilities	
Accounts payable	$ 50,000.00
Income taxes payable	50,000.00
Total current liabilities	$ 100,000.00
Long Term Liabilities	
Bonds payable	$ 1,000,000.00
Total liabilities	$ 1,100,000.00
Owner's Equity	
Capital stock—common	$ 2,000,000.00
Paid in capital in excess of par—common stock	500,000.00
Total contributed capital	$ 2,500,000.00
Retained earnings	$ 5,315,000.00
Total owner's equity	$ 7,815,000.00
Total liabilities and owner's equity	$ 8,915,000.00

NOTE:

Applying the accounting formula to the statement of financial position, the total assets of $8,915,000.00 equal the total liabilities of $1,100,000.00 plus owners equity of $7,815,000.00. Or total assets of $8,915,000.00 less total liabilities of $1,100,000.00 equal total owner's equity of $7,815,000.00. Had the statement of financial position been presented in another form, all the asset accounts would have been listed on the left side and all of the liability and owner's equity accounts would have been listed on the right side. This alternative presentation would have been consistent with the accounting procedure of debits on the left and credits on the right and with the basic accounting formula. Asset accounts are increased by debits and would appear on the left side. Conversely, liability and owner's equity accounts would appear on the right side and they are increased by credits and are classified as credit accounts.

The statement of financial position also illustrates the residual nature of owner's equity. The total assets of $8,915,000.00 less the total liabilities of $1,100,000.00 equals the owner's equity of $7,815,000.00. Therefore, the owner's share of the firm is mostly noncash assets such as marketable securities, accounts receivable, inventories, land, and plant and equipment. Hence the firm might have difficulty if required to liquidate its liabilities immediately. It would have only $10,000.00 in cash to pay out immediately, and if the other assets could not be sold for at least their book value in such a liquidation, the shareholders would not receive the amounts indicated in the statement of financial position.

QUESTIONS

1. What is the role of accounting in the business firm?
2. What is the purpose of the accounting concept of profit?
3. How are the accounting formula and the double entry system related?
4. How are asset values increased? Or decreased?
5. How are liability and owner's equity account values increased? Or decreased?
6. What is the nature of the owner's equity interest in a firm?
7. What might a typical owner's equity section of a corporate statement of financial position look like?
8. Define accounting pluses and minuses.
9. What is the accounting book of original entry?
10. What is the function of the accounting book of original entry?
11. Describe the flow of accounting information.
12. How can access to the general ledger be controlled?
13. Give at least three examples of accounting journal entries and the events that caused them.
14. Describe the process of posting to the general ledger.
15. What is a trial balance and what is its purpose?

REFERENCES

Anthony, Robert N., and James S. Reece, *Management Accounting* (Homewood, Illinois: Richard D. Irwin, 1975).

Chorba, George J., *Accounting for Managers* (New York: American Management Association Extension Institute, 1978).

Gordon, Myron J., and Gordon Shillinglaw, *Accounting; A Management Approach* (Homewood, Illinois: Richard D. Irwin, 1974).

Kieso, Donald E., and Jerry J. Weygandt, *Intermediate Accounting* (New York: John Wiley & Sons, 1977).

CHAPTER 2

ACCOUNTING CONTROLS AND PROCEDURES

As discussed in the preceding chapter, control over journal entries is required. Despite these controls, it is also necessary to control the posting of the journal entries to the general ledger. The purpose of controlling the postings to the general ledger is to prevent the direct posting of false entries. It is insufficient to control the journal entries; the posting must also be controlled or it is possible for someone to make a direct unauthorized posting which would nullify journal entry controls.

One method to control posting is to require a reference to the source of the posting, that is, the journal and the page within the journal from which the posting was made. Another control is to assign the responsibility for posting the general ledger to one group. This prevents the dispersion of responsibility and limits access to the general ledger. In addition, on a sampling basis, journal entries should be traced to the journal entry control to verify that it is an authorized entry. A sample of postings can also be taken by internal audit or the general ledger supervisor and traced back to the journal to verify authenticity. When the posting of the journal entries has been satisfactorily completed, the next step is to verify that the general ledger is in balance.

PRELIMINARY TRIAL BALANCES

The purpose of a preliminary trial balance is to determine whether the general ledger is in balance after posting from the journals. At this point all journal entries have been posted except the adjusting and closing entries. In addition to verifying whether the debits and credits are in balance, a comparison with past years may reveal areas where it is necessary to investigate further.

The preliminary trial balance is prepared by listing all accounts that have balances in the general ledger and listing them in general ledger order. After the posting is complete the debits and credits in each account in the general ledger are totaled and a balance for each account is calculated. Then the accounts are listed in the order in which they appear in the general ledger. The balance in each account determines whether it is listed as a debit or credit.

Accounts are usually listed in the general ledger in the order of the statement of financial position, that is, the assets are listed first, the liability accounts are listed next, then the owner's equity accounts are listed, and finally the income and expense accounts. Within the order of listing all contraaccounts are included with the asset and liability accounts to which they apply. When the listing of each account number, the account description, and the amount is complete, the debits and credits are totaled to verify that the general ledger is in balance.

If the trial balance does not balance, the listings must be checked to the general ledger to verify whether the accounts and their balances were listed correctly. If the listing is correct, each account in the general ledger must be rebalanced. Assuming that the balances are correct, each posting must be retraced to the journals. Once the error or errors have been found and the preliminary trial balance balanced, the end of period accounting adjustments can be recorded.[1]

THE END OF ACCOUNTING PERIOD ADJUSTMENTS

The end of accounting period adjustments include accruals, the expiration of prepaid expenses, and miscellaneous adjustments. To ensure the accuracy of the financial statements, the end of period adjustments include all previously unrecorded transactions.

DIRECT LABOR ACCRUALS

The payday for direct factory employees may not coincide with the end of the accounting period. Therefore, the portion of the pay period applicable to the accounting period being closed must be included in the financial statements. If accurate information is not available the

---

[1]Donald E. Kieso and Jerry J. Weygandt, *Intermediate Accounting* (New York: John Wiley & Sons, 1977), pp. 59–96.

accrual can be estimated. Techniques for preparing an accrual are: (*a*) if information about the pay period to be accrued is not available, a percent based on the number of days to be accrued versus the total days in the pay period can be applied to the prior pay period total dollars; (*b*) if the total pay for the period to be accrued is available, the percentage computed in (*a*) can be applied to the total dollars for the pay period; (*c*) if the accrual can be prepared late enough in the following accounting period, the actual amounts applicable to the period of the accrual are available; and (*d*) the accrual can be estimated by computing an average direct labor rate and applying that rate to the estimated hours to be accrued. Regardless of which method is used, however, direct labor must be accrued at the end of the accounting period to record an accurate period cost and the applicable liabilities. An example of an accrual entry for direct labor is:

19XX

Dec. 31	Work in process—direct labor	$XXX	
	Accrued payroll payable		$XXX
	To accrue XX days of direct labor.		

ACCRUAL OF INDIRECT LABOR

Indirect labor must also be accrued at the end of the accounting period. The problem is the same as with direct labor: the pay period and the accounting period may not coincide. The accrual of indirect labor, however, is more complex. The complexity is caused by the large number of accounts that indirect labor may be charged to. For example, indirect labor consists of foreman's clerks, security personnel, materials handlers, receiving and inspection personnel, foreman, and millwrights. The accrual must be estimated to include the amounts applicable to each account. The same estimating procedures that are used for direct labor can be used for indirect labor except the amounts must be spread among more accounts. An example of an accrual entry for indirect labor is:

19XX

Dec. 31	Factory overhead—clerks' salaries	$XXX
	Factory overhead—security salaries	$XXX
	Factory overhead—material handlers' salaries	$XXX
	Factory overhead—receiving and inspection salaries	$XXX
	Factory overhead—foremen's salaries	$XXX

Factory overhead—millwrights'
salaries $XXX
   Accrued payroll payable $XXX
To accrue XX days of indirect labor.

ACCRUAL OF MATERIAL

Because of the time it takes to process vendor's invoices and to match receipts and invoices, it is necessary to prepare an accrual journal for direct and indirect material at the end of the accounting period. Usually there are criteria for the amounts of invoices and receivers that are considered material and should be accrued. For example, $500.00 may be the minimum amount considered material for a financial closing at the end of a month, while $100.00 may be considered the minimum amount for the end of the year financial closing. Any invoices or receiving slips that equal or exceed the specified criteria are included in the accrual. Material accruals usually take three forms. (a) Invoices that are received and not vouchered are those that have been received and matched with receiving slips. The invoices and receiving slips have also been matched with the purchase order but because the invoices have not been vouchered their effect on assets, expense, and liabilities is not recorded; therefore they must be accrued. (b) For material that has been received but has not been vouchered only internal documentation, the receiving slip, is available. The vendor's invoice has not been received; therefore, the voucher cannot be prepared. But it is necessary to record the effect of these purchases on the firm's assets, liabilities, and expenses. The receiving slips are checked against the purchase order, the account number to be charged recorded on the receiving slip, the value of the receiver calculated, and, if it meets the monetary criteria, the value of the receiving slip is included in the accrual journal entry. (c) For goods-in-transit invoices have been received from vendors but the material has not arrived. Whether an invoice is included in the accrual depends on the accrual criteria discussed above and the FOB (free on board) point. If the material was shipped FOB the vendor's plant, the material is the property of the purchaser while in transit. Therefore, if it was shipped before the end of the accounting period it should be included in the accrual. Conversely, if the FOB point is the purchaser's plant the invoice should not be included in the accrual. The journal entry for the material accrual is:

19XX
Dec. 31   Raw materials inventory $XXX

Manufacturing indirect inventories	$XXX
Maintenance expense	$XXX
Office supplies expense	$XXX
Fuel oil expense	$XXX
Accrued liabilities	$XXX

To record direct and indirect materials accrued at the end of the accounting period.

RECORDING OF EXPIRED INSURANCE AND MISCELLANEOUS ADJUSTMENTS

Another necessary entry at the end of the accounting period is the recording of the insurance expense and other miscellaneous adjustments. When the premium is paid for insurance, the period covered is usually two or three years in advance. As the period covered by the advance payment expires, the actual insurance expense must be recorded. The entry for the advance is:

19XX
Dec. 31

Unexpired insurance	$XXX
Vouchers payable	$XXX

To record the payment of the insurance premium in advance.

The entry to record the expiration of the insurance coverage and the resulting insurance expense is:

19XX
Dec. 31

Insurance expense	$XXX
Unexpired insurance	$XXX

To record the expiration of insurance coverage and the resulting insurance expense.

The miscellaneous adjustments include a number of different adjustments such as any additional goods-in-transit accrual, the recording of the supplies inventory, the adjustment of the finished goods, work in process, and raw materials inventories to market and then to dollar value LIFO, the adjustment of all inventories to the physical counts, the accrual of sales for contracts accounted for on the percentage of completion basis, and the adjustment of the income statement for state and federal income taxes. Because of the various sources for material, shipments that should be accrued within the financial closing are not received until well into the next fiscal year. Therefore, as the last entries are prepared, any additional goods-in-transit accrual should be recorded. The journal entry to record the additional goods-in-transit is similar to the entry prepared for the original goods-in-transit accrual discussed above.

The maintenance of a detailed day to day record of the transactions affecting the supply inventory is too expensive to cost justify. Usually then, in place of a perpetual inventory record, a physical inventory is made of the supplies on hand at the end of the fiscal year. The results of the inventory are journalized to establish the supplies inventory at the end of the fiscal year. The journal entry to record the supplies inventory at fiscal year end is:

```
19XX
Dec. 31   Supplies inventory                      $XXX
             Supplies expense                              $XXX
          To record the supplies inventory based
          on a physical inventory at fiscal year
          end.
```

Most firms including construction firms, record their inventories at cost or some variation of cost for internal use. However, for outside reporting the inventories must be reported at the lower of cost or market value and some firms use dollar value LIFO to report externally. An adjustment must be made to change the inventory values from those used internally to those used for external reporting. Because the values used internally are to be retained for future use and any change would require an adjustment of the subsidiary records, that is, all inventory detail records, the entries are not made directly to the inventory accounts but to contrainventory (asset) accounts. A typical entry to adjust finished goods inventory to market is:

```
19XX
Dec. 31   Loss to adjust finished goods inventory
             to market                            $XXX
             Allowance for loss to adjust finished
             goods inventory to market                    $XXX
          To record the adjustment of the
          finished goods inventory to the lower of
          cost or market.
```

The journal entry to record the adjustment of the finished goods inventory to dollar value LIFO is:

19XX

Dec. 31   Loss caused by adjustment of finished
                goods inventory to dollar value LIFO $XXX
                    Allowance for loss to adjust finished
                    goods inventory to dollar value LIFO $XXX
                To record the adjustment of the
                finished goods inventory value to dollar
                value LIFO.

One requirement of the accounting profession is that all inventories be physically counted at least once a year. Because of the complexities inherent in physical inventories the recorded inventory and the physical inventory will not agree. Therefore, the inventory value must be adjusted to that established by the physical count. An example of a journal entry to adjust the inventory carrying value to the physical is:

19XX

Dec. 31   Loss to adjust finished goods inventory
                to the physical count $XXX
                    Allowance for loss to adjust finished
                    goods inventory to the physical count $XXX
                To adjust the finished goods inventory
                value to the results of the physical
                inventory of Dec. 20, 19XX.

The physical count could result in a gain or loss. In this example it was assumed that the count resulted in a loss. Additionally, the adjusting entry could have been made to the inventory account rather than to an allowance account. Because the physical counts can be over as well as under the carrying value of the inventory, some accountants prefer to use the allowance account rather than continually adjust the inventory account.

If a construction firm is using the percentage completion method for recording work done on contracts that are in progress, an entry is necessary to record the effects of the work completed and not recorded through the processing of an invoice. Both the sales and costs for each contract must be recorded. Although the methods for recording unbilled sales are discussed later in this book, let us mention three methods that may be used: (a) engineering studies may be made to determine the actual work completed compared to the total work required by the contract, (b) an engineering estimate may be made of the work completed relative to the total required, and (*c*) an estimate may be made based on the percent of cost incurred compared to the total estimated cost in the contract. An example of the journal entry required to record the sales not billed on construction contracts is:

19XX
Dec. 31   Accrued accounts receivable—
      construction contracts $XXX
        Accrued sales—construction contracts $XXX
      To record unbilled sales on construction
      contracts recorded by the percentage of
      completion method.

19XX
Dec. 31   Accrued cost of sales—subcontracts $XXX
        Accrued accounts payable—
      construction contracts $XXX
      To record unpaid subcontracting costs
      for construction contracts recorded by
      the percentage of completion method.

One adjustment that must be made before the preparation of financial statements is the adjustment for subdivision, state, and federal income taxes. Because these taxes are based on net income before income taxes, they cannot be computed until all other adjustments have been made. An example of the adjusting entry for income taxes is:

19XX
Dec. 31   Subdivision income tax expense $XXX
      State income tax expense $XXX
      Federal income tax expense $XXX
        Subdivision income taxes payable $XXX
        State income taxes payable $XXX
        Federal income taxes payable $XXX
      To record the expense and related
      liabilities for income taxes at fiscal
      year end.

THE EFFECT OF ADJUSTMENTS ON NET INCOME AND FINANCIAL RATIOS

To reflect a reasonably accurate net income all expenses and revenues must be either recognized and recorded or accrued. The effect on net income of a charge to an expense account whether it is properly vouchered or only accrued is the same. The charge to expense reduces net income. Conversely, revenue, whether accrued or recorded in the sales journal, increases net income. Because the effect of either method of recording is the same, it is prudent and necessary to record all accruals.

The effect of the accrual of asset and liability account charges is more subtle than the effects of expense and income charges. Even if the material accrued did not affect any income or expense accounts, it is still important that the accrual be recorded because of its effect on the financial ratios, especially the current ratio. Assume that we have a current ratio of 2 to 1, that is, there are $2 of current assets for every $1 of current liabilities, which is computed by dividing the total current assets by the total of the current liabilities. If the material accrual results in a debit to inventory and a credit to the accrued liability account on a 1 to 1 basis, the entry would tend to reduce the current ratio below 2 to 1. If we assume the accrual of material reduced the current ratio from 2 to 1 to 1.5 or 1.75 to 1, the financial statements would be misleading without recording the accrual.

THE CLOSING ENTRIES

The closing entries are necessary to remove all balances from the temporary accounts so they can be used in the next accounting period and to transfer the effect of closing the temporary accounts to the proper owner's equity account. The closing entries are designed to offset expense and income accounts and identify the residual amount which is transferred to a permanent account, retained earnings. Expense accounts are closed by crediting each account (assuming that expense accounts have a debit balance, which is normally the case) and debiting an income summary account that is set up for this purpose. All income accounts are debited and the income summary account is credited. After these entries are completed, the income summary has a credit balance (assuming a net income and not a net loss) that is equal to the net income for the period. This balance is transferred through a journal entry to the retained earnings account. The closing entries take the following form:

19XX

Dec. 31	Income summary account	$XXX	
	Salaries expense		$XXX
	Insurance expense		$XXX

Supplies expense $XXX
To close the expense accounts to the
income summary account at the end of
the fiscal year.

19XX
Dec. 31 Sales $XXX
Miscellaneous income $XXX
Income summary account $XXX
To close the revenue accounts to the
income summary account at the end of
the fiscal year.

19XX
Dec. 31 Income summary account $XXX
Retained earnings $XXX
To close the income summary account
and record the net income after taxes
for the fiscal year.

THE POST CLOSING TRIAL BALANCE

The post closing trial balance is a listing of the open accounts in the general ledger after the closing entries have been posted. If the journal entries have been posted correctly the only accounts with balances in the general ledger are the permanent asset and liability accounts. When the listing is complete no temporary accounts— expense or income accounts—should be included. None of the temporary accounts should have any remaining balances. In addition, the debit and credit balances in the post closing trial balance should be equal.

If the debit and credit balances are not equal or there are temporary account balances included in the post closing trial balance, the adjusting and closing entries must be checked. The recordable events must be retraced through the adjusting and closing journal entries. The posting of the journal entries must be followed through the general ledger and the error corrected. Once the post closing trial balance is correct, the financial statements can be prepared. An example of a post closing trial balance is included as Exhibit 2.

THE PREPARATION OF THE FINANCIAL STATEMENTS

The post closing trial balance and the closing entries are the basis for the preparation of the financial statements. The balance sheet for the current period can be prepared directly from the post closing trial balance. The accounts in the post closing trial balance must be rearranged into correct balance sheet classifications. Any comparative balance sheets (statements of financial position) must be added to the current period financial statements.

The financial statements required under present rules (not including the recent requirement for inflation adjusted income statements and the supplementary disclosures) are: (a) the statement of financial position, (b) the income statement, (c) the statement of changes in financial position (required whenever an income statement is included in the financial statements), and (d) the statement of retained earnings. As noted above, the statement of financial position (balance sheet) can be prepared directly from the post closing trial balance. The statement of changes in financial position can be prepared from the current and the prior year's statement of financial position.

The statement of retained earnings is also prepared from the post closing trial balance except for the balance at the beginning of the period. The preparation of the income statement is more complex. It is necessary to use the closing entries and some analysis of accounts to prepare the details required for the income statement. Assuming that there is a statement of cost of goods sold required, the beginning and ending inventory balances for raw materials, work in process, and finished goods are taken from the current and prior period statement of financial position. The purchases and details for the direct labor and overhead are listed from account analysis. The statements usually include a detailed listing of the individual overhead accounts. For construction firms the general statement preparation procedures are the same except that the statement of cost of sales is substituted for the statement of cost of goods sold.

The remaining balances required for the income statement are taken from the closing entries. The expense and income accounts are listed in acceptable order and the net income should equal the closing entry to the retained earnings account. Any comparative data can be added from last year's income statement and cost of goods sold or cost of sales.

THE CHART OF ACCOUNTS

The chart of accounts is the record of the financial organization of the firm. Preparation of the chart of accounts requires judgment. The list of accounts must be detailed enough to be useful for analyzing and managing the firm without creating excessive detail and confusion.

CLASSIFICATION OF THE ACCOUNTS

Accounts are classified within two broad categories: (*a*) statement of financial position accounts (balance sheet accounts) and (*b*) expense and income accounts which measure the change in the retained earnings section of the owner's equity from period to period. Within these two categories the accounts are classified as assets, liabilities, and owner's equity accounts and as direct project expense, indirect project expense, and central expense accounts.

The asset accounts are further classified as current assets, operational assets—tangible, operational assets—intangible, and other assets.[2] The liability accounts are classified as current liabilities, long term liabilities, and deferred credits. Within owner's equity the accounts are classified as common stock, preferred stock, paid in capital in excess of par, retained earnings, and in some cases treasury stock.

Direct project expenses include labor used directly on a project and any material or subcontracting used directly on the project. There is also a subclassification for equipment expense—direct to accumulate the cost of the use of equipment on the project. Indirect project expenses include indirect expenses for one project that must be classified as indirect because they apply to more than one item within the project and indirect expenses that apply to more than one project and therefore must be allocated. Central expenses are separately accumulated because they are expenses of the home or central office and include costs that must be allocated over the total business of the firm. They include such costs as central administration, contract administration, and financing costs. The revenues are classified into contract revenues, by contract and miscellaneous or other revenues.

The classification of accounts discussed above is compatible with the classification of accounts as real or nominal. Real accounts are accounts that remain open from period to period. These accounts are the accounts that reflect the balances as of a certain date. The real accounts are the accounts that are listed in the statement of financial position. The nominal accounts are those accounts that are closed at the end of each accounting period. These are the accounts

[2]Glenn A. Welsch, Charles T. Zlatkovich, and Walter T. Harrison, Jr., *Intermediate Accounting* (Homewood, Illinois: Richard D. Irwin, 1979), p. 137.

that are listed in the income statement. These accounts reflect the change in owner's equity (retained earnings) for the period. The chart of accounts is organized and numbered to separate the real and nominal accounts.

SELECTION OF THE CHART OF ACCOUNTS

Usually all accounts to be included in the chart of accounts cannot be selected when the chart of accounts is originally prepared. The chart of accounts is prepared under the assumption of growth, therefore room must be left for expansion. The number of accounts is selected on the basis of at least two criteria: (a) the number of accounts must be large enough to provide valid information for managing the firm without the confusion of too much detail, and (b) there must be sufficient number of accounts to assure users that the information disclosed in the financial statements is not misleading.

Generally, the objectives in selecting the number of accounts are to provide enough information to materially represent the results of operations for the period covered by the financial statements and to support consistent comparison of operating results from period to period. The chart of accounts must be flexible but it must not lead to material inconsistencies from period to period comparisons.

Account titles must be selected with care. The titles must describe what is included in the accounts in terms that are general enough not to require an account for each entry and yet not so general that the title is meaningless. For example, the title cash may not be sufficient in some cases. The cash account may need to be separated into demand deposit cash—detailed by bank, time deposit cash—detailed by bank, restricted cash balances (such as compensating balances)—detailed by bank, and imprest fund cash.

Also, there should be available a description of what kind of charges to include within each account. If such a description is not available the accuracy of the account titles and the need for consistency are compromised. The account description needs to clearly tell the user which costs are to be included in this account. An example of an account description is one that could be used for project direct labor—the labor that is used directly on the objective of the project. It includes labor for constructing a building, the laying of a roadway, the installation of air conditioning, or the construction of a bridge. This account is not to include any charges for support personnel such as security personnel, clerks, or randomly assigned truck drivers. Heavy equipment operators are included in this account to the extent they are assigned to one project or one item within a project. The criterion for assigning cost to this account is whether the employee is directly assigned to one project or item within a project for at least four or more hours in any day. This account is classified as a nominal account and appears on the income statement under the subclassification of project direct labor.

NUMBERING OF THE ACCOUNTS

There was a time when the accounts carried only a description. But in the modern world of the computer account titles are awkward and less efficient to use than digits. Therefore, the chart of accounts is also numbered. The accounts are not randomly numbered, however. The account numbers are assigned in an orderly sequence. The numbering system follows the financial statements and their sub-classifications. The numbers are assigned first to the asset accounts, then to the liability and owner's equity accounts. Last, the income and expense accounts are assigned numbers. Within these categories the digits are assigned by current assets, operational assets, and other assets and in a similar manner for liabilities and owner's equity.

The numbering assignments must also leave room for expansion. Therefore, the numbers are assigned in blocks or groups. For example, 10000 may be assigned for assets, 20000 for liabilities, 30000 for owner's equity, 40000 for income accounts, 50000 for expense accounts, and 60000 for the income summary account. Within the 10000 group a block is assigned for current assets; for example, current assets begin with the series 11000, the current asset demand deposits may be numbered 11001. The tangible operational assets are assigned 12000, the intangible operational assets 13000, and the other assets 14000. The current liabilities could be numbered 21000, the long term liabilities 22000, and the deferred credits 23000.

The owner's equity accounts are usually numbered 30000 with the capital accounts as 31000, paid in capital as 32000, dividends 33000, retained earnings 34000, and treasury stock 35000. Because the income and expense accounts measure the change in owner's equity, they are, if possible included in the same series as the other owner's equity accounts. Usually, however, the expense and income accounts are so numerous that separate series are assigned for each. The number series assigned to income and expense accounts are referenced and programmed to be summarized as detail accounts for retained earnings. As an example, the income accounts are assigned the series 40000 and the expense accounts 50000. Exhibit 3 is an example of a chart of accounts.

CONTROL OF CHANGES TO THE CHART OF ACCOUNTS

Any changes to the chart of accounts, including the account descriptions, must be strictly controlled. Requests for additions to the chart of accounts or for deletions should be submitted in writing. Also, any changes in the account descriptions should be written requests. The requests should be submitted to one person who is charged with responsibility for these changes. The employee responsible discusses the proposed changes with the chief financial officer. If the changes are warranted, the formal changes are prepared in writing by the responsible employee. The written requests must then have the approval of the responsible officer. The need for control of changes in account information is to prevent compromise of assets and ensure comparability of financial information for internal and external use.

THE UNIQUE PROBLEM OF FINANCIAL CONTROL IN THE CONSTRUCTION INDUSTRY

Financial control in the construction industry is more difficult than in almost any other industry. One of the major control problems is the lack of locational control. In manufacturing and almost all other industries, operations are included under one roof or within a number of plant or branch locations. However, construction sites are usually widely dispersed and each is unique and unlike all others. The geographical locations of the projects may also be widely separated, thus adding to the control problems.

To keep the costs of controls within reasonable amounts, financial control systems rely heavily on the construction field personnel. All of these factors make it difficult to install, test, and control reporting systems. Remember that field personnel are usually selected because of their ability to complete a project within the contract physical specifications and time constraints. They are not necessarily experts on or concerned with financial and accounting controls. Because of the possible weaknesses in control systems in the construction industry, accounting and financial controls must be specifically and carefully designed and strictly adhered to. Communication with field personnel must be continuous.

SUMMARY

This chapter has discussed the closing procedures at the end of the accounting period, the chart of accounts, the need for control of both procedures, and the unique control problems inherent in the construction industry. The closing procedures were traced from the posting to the general ledger through the preliminary trial balance, the end of period accounting adjustments, the closing accounting journal entries, the post closing trial balance, and the preparation of financial statements at the end of the accounting period.

The chart of accounts was defined and illustrated. The classification of accounts was demonstrated and real and nominal accounts were defined. The importance of the chart of accounts was emphasized including the selection of the number of accounts, the proper titling of the accounts, and the need for an accurate and complete description of the nature of the costs to be charged to each account. Because of the growth in the use of computers, the numbering of the chart of accounts was commented upon. In addition, the need for the control and approval of any changes to the content of the chart of accounts was stressed.

EXHIBIT 2.   Post Closing Trial Balance at December 31, 19XX

Description	Debit	Credit
Cash	$10,000	
Marketable securities	100,000	
Land	1,000,000	
Buildings	10,000,000	
Allowance for depreciation buildings		$2,000,000
Machinery and equipment	20,000,000	
Allowance for depreciation machinery and equipment		5,000,000
Patents	200,000	
Unexpired insurance	5,000	
Trade accounts payable		7,000,000
Notes payable		5,000,000
Bonds payable		10,000,000
Discount on bonds payable	10,000	
Capital stock—common		1,000,000
Capital stock—preferred		500,000
Paid in capital in excess of par—common		200,000
Retained earnings		645,000
Treasury stock—common	20,000	
Total	$31,345,000	$31,345,000

EXHIBIT 3. Chart of Accounts

Description	Account Number
Demand deposits	11001
Time deposits	11002
Treasury bills	11004
Certificates of deposit	11005
Land	12001
Buildings	12003
Allowance for depreciation buildings	12004
Machinery and equipment	12006
Allowance for depreciation machinery and equipment	12007
Patents	13001
Unexpired insurance	14001
Trade accounts payable	21001
Notes payable	22001
Bonds payable	22004
Discount on bonds payable	22005
Capital stock—common	31001
Capital stock—preferred	31002
Paid in capital in excess of par—common	32001
Retained earnings	34001
Treasury stock—common	35001
Sales	41001
Sales returns and allowances	41002
Sales discounts	41003
Interest income	42001
Miscellaneous income	43001
Raw material used	51001
Direct labor expense	52001
Indirect factory labor expense	53001
Depreciation expense	54001
Utilities expense	55001
Supplies expense	56001
Sales salaries expense	57001
General and administration salaries expense	58001
Sales travel expense	59001
General and administrative travel expense	51101
General and administrative depreciation expense	51201
Consulting costs	51301
Stationary and office supplies	51401

QUESTIONS

1. How are postings to the general ledger controlled?
2. List and describe the end of period accounting adjustments.
3. What is the post closing trial balance and what is the purpose of a trial balance in general?
4. Describe and give examples of the closing entries.
5. How are financial statements prepared?
6. In detail, describe and discuss the chart of accounts.
7. What are "real" accounts?
8. What are "nominal" accounts?
9. Of what importance is the chart of accounts?
10. Why and how are changes to the chart of accounts controlled?
11. Discuss and describe, in detail, the unique problems of financial control in the construction industry.

REFERENCES

Anthony, Robert N., and James S. Reece, *Management Accounting* (Homewood, Illinois: Richard D. Irwin, 1975).

Chorba, George J., *Accounting For Managers* (New York: American Management Association Extension Institute, 1978).

Gordon, Myron J., and Gordon Shillinglaw, *Accounting: A Management Approach* (Homewood, Illinois: Richard D. Irwin, 1974).

Kieso, Donald E., and Jerry J. Weygandt, *Intermediate Accounting* (New York: John Wiley & Sons, 1977).

Montgomery, A. Thompson, *Managerial Accounting Information* (Menlo Park, California: Addison-Wesley Publishing Company, 1979).

Welsch, Glenn A., Charles T. Zlatkovich, and Walter T. Harrison, Jr., *Intermediate Accounting* (Homewood, Illinois: Richard D. Irwin, 1979).

CHAPTER 3

ACCOUNTING PRINCIPLES AND POLICIES

Generally accepted accounting principles guide the presentation of almost all financial information. These principles are usually determined by the accounting profession itself. The nature of these principles, who is involved in the process of determining the principles, and who interprets the application of these principles are the subjects of this chapter.

GENERALLY ACCEPTED ACCOUNTING PRINCIPLES (GAAP)

The generally accepted accounting principles have developed over a long history of practice mainly through trial and error. Their purpose is to provide a framework within which financial information may be presented. This framework provides some similarity and continuity in the publication of financial information. A framework of principles should make the information more meaningful and useful to those who own the firm as well as those interested in evaluating the firm as a potential supplier of funds, as a customer, or even as a potential regulator.

THE MATCHING PRINCIPLE

Most financial statements are prepared under the principle of matching. The purpose of matching is to associate the costs of a firm with the revenue the costs produced. For example, if a product is manufactured in one period and the costs of manufacture recorded within the period of manufacture as an operating cost, a loss would be reported if the product were sold in the next period. There would not be any matching. If the firm manufactured a large number of products and sold products in large quantities it would be almost impossible to evaluate the firm's performance. The matching principle is, for example, responsible for the development of accrual systems and for the use of the asset called inventory. Under the cash basis of accounting the revenue and cost of operations were difficult to match. Labor usually must be paid immediately, but there is often a long collection cycle before money is received from customers. This problem led to the accrual system in which both costs and sales are accrued to match the cost and revenue.

In addition, the sale of products and services in periods after manufacture caused the use of inventory as a method of holding the cost until the product is sold. The cost of the product manufactured is put in the balance sheet as an asset until the product is sold and then the cost is transferred to the income statement and deducted from the sales revenue to compute net income or loss. The inventory is treated as an asset because in the future it will result in an increase in the value of the firm through its sale unless it is sold at a price that is below total costs. Matching supports the evaluation of the firm by equating the costs of a product with the revenue from its sale.

THE PRINCIPLE OF CONSISTENCY (COMPARABILITY)

The purpose of consistency is to encourage the comparability of financial information from period to period and firm to firm. The principle of consistency prevents a firm from arbitrarily switching from one depreciation method to another within or between accounting periods. If it were not for the consistency requirement firms could switch from one inventory cost flow assumption to another and cause its financial statements to be confusing. The principle of consistency does not prevent an entity from changing its accounting practices but there must be substantial reasons for the change and the change and its effect must be disclosed.

The intent of the principle of consistency is to keep financial information on a comparable basis and allow the evaluation of the financial results by comparison of past and present periods which are useful to identify changes and trends. The consistency principle has been one of the reasons the accounting profession has been reluctant to support the preparation of inflation adjusted accounting statements. Because inflation changes the basis for the financial statements the degree of comparability and consistency that remains is questionable.

THE PRINCIPLE OF FULL DISCLOSURE

The purpose of the principle of full disclosure is to assure the user of financial information that all relevant information has been disclosed, preferably without confusing the user. This principle prevents the partial disclosure of financial information that could mislead the user of the information. If, for example, the choice were to reveal or not to reveal a relationship between the preparer and the reviewer of financial information, the principle of full disclosure would require that the relationship and its nature be revealed. When there is a choice of what to disclose and of the extent to which information should be disclosed, the alternative should be chosen that results in the fuller disclosure.

OBJECTIVITY (VERIFIABILITY) OF FINANCIAL INFORMATION

Financial information should be as objective as possible. Users would not be able to rely on financial information if it were compiled and presented as strictly the preparer's interpretation. The information must be more objective than to present just the preparer's views. This principle does not mean that all financial information would be presented in the same way or in the same amounts regardless of who prepared it. The objectivity principle was designed to assure the user that the financial information is prepared in a manner that enables a reasonable person to verify that the amounts and presentation are within the permitted alternatives. A reasonable person acting as a reviewer should be able to reach the same conclusions as the preparer.

THE PRINCIPLE OF MATERIALITY

In one sense, materiality is similar to significance. However, it is possible for an amount to be significant but to not have a material effect on the interpretation of the financial results. The principle of materiality implies that the inclusion of this information would effect the user's interpretation of that information. The applicability of other accounting principles and concepts is modified by the principle of materiality. The other principles and concepts are only applied to financial information that is material in nature. That is, for example, there is no need to disclose information if the effect of the disclosure is not material. One important difficulty with the principle of materiality is the interpretation of what is material. Materiality depends on the judgment of the preparer and reviewer of financial information.

Although some accounting theorists have suggested that 10% be adopted as the criterion for materiality, the accounting profession has not yet followed this suggestion. What the accounting theorists were attempting was to limit the area of judgment and thus the divergent views concerning what is material. For example, if the disclosure of an expense were being debated, disclosure would be required if the expense affected reported net income or loss by approximately 10% or more.

The question of materiality is also complicated by the need to decide whether the decision should be based on a single amount or the accumulation of a number of amounts or items. The professional standards generally require that the cumulative effect be considered. However, this is again a matter of judgment to be applied by the preparer and the reviewer of financial information. Materiality is a criterion for the preparation of financial information but it is difficult to determine objectively.

CONSERVATISM: A CONTROVERSIAL PRINCIPLE

The intent of the principle of conservatism is to prevent the overstatement of financial information. Many accountants believe that a projection or an estimate of gains as well as losses, if applicable and if reliable, should be included in published financial information. The principle of conservatism, however, does not support this approach.

Conservatism requires that the preparer and reviewer of financial information choose the alternative which will be least likely to overstate assets and net income. This principle results in the choice of the alternative with the lowest probability of overstatement. It in effect says: "Recognize (anticipate) all losses but only record gains when they actually occur." Even if the preparer and reviewer are almost certain that a gain will occur, it cannot be included until it actually happens. The principle of conservatism attempts to preclude overstatement of financial information even if the overstatement results only from optimism rather than an attempt to mislead.

THE ECONOMIC ENTITY ASSUMPTION

The economic entity assumption presumes that the performance of the firm (entity) can be measured separately and apart from the performance of all other firms (entities). It assumes that the entity is a separate economic unit and its performance can and should be measured.

THE ASSUMPTION OF THE GOING CONCERN

This assumption has had an extensive effect on accounting theory and practice. When a preparer or reviewer evaluates financial information it is always from the perspective that the firm will continue in operation and is not about to liquidate. This assumption has affected the emphasis placed on the different financial statements and the values used to report items and rights owned by the firm. The assumption that the firm will continue in business has led to concentration on the financial statement that measures performance, that is, the income statement. Because the firm is viewed as a continuum, the income statement, which measures performance, becomes the statement that is emphasized and considered the most important. Continuity accents performance rather than stewardship.

In the past, one of the prime reasons given for the use of historical cost for asset valuations was that the true value is the use of the assets by the firm and that any value they have on disposal is only relevant if liquidation is contemplated. If the firm is to continue in business, the exit values of its assets are not relevant. In addition, these values are reported in the least important of the two primary financial statements. Currently, the ability of the balance sheet to provide security for debt is becoming more important.

THE MONETARY UNIT ASSUMPTION

The monetary unit assumption presumes that all financial transactions can be recorded in the prevailing units of money. That is, the transactions that apply can be expressed in monetary terms and the face value of the monetary unit can be used to value the necessary transactions. This assumption is applicable even though the monetary unit may not remain stable in value. The monetary unit may need to be adjusted for general inflation or a change in its underlying value but it is still the basis for recording financial transactions.

THE ASSUMPTION OF PERIODICITY

This assumption states that the accounting and financial cycle of a firm or entity can be broken into distinct periods. The true value of a firm cannot be measured accurately until the firm is liquidated. In some cases that may never occur. Therefore, the assumption is that a firm's life can be broken into distinct periods of time for measuring performance. Within the United States it is almost universal practice to use a fiscal year of 12 consecutive months. This has been adopted as universal practice despite the general rule of accounting which is: The fiscal period may be one year or the operating cycle whichever is longer.

THE HISTORICAL COST ASSUMPTION

The historical cost assumption presumes that assets will be recorded at cost. Cost will continue to be the carrying value of the firm's assets until they are exchanged or there is a quasireorganization of the firm. This assumption causes assets to remain on the firm's records at historical cost, and their value cannot be changed to a more current amount whether that current value be market, replacement, or some other value. A recent pronouncement of the FASB requires that certain asset values be modified to reflect inflationary changes and replacement costs but this information does not replace historical cost as the primary method of valuation but is supplementary information to the historical cost values. The only current requirement for modification of historical cost is the lower of cost or market rule which is discussed later in the book.

THE PRINCIPLE OF REVENUE REALIZATION

To provide a reasonable amount of comparability and to prevent financial statements from being misleading, the accounting profession developed a guide for the recognition of revenue. The recognition of revenue cannot take place until both the substantial completion of the earnings process and exchange have occurred, which is generally interpreted to mean there has been a sales transaction. The exchange is usually evidenced by the transfer of ownership of a good or the transfer of consideration for goods or services.

   Other alternatives have been suggested to replace the current principle of revenue recognition. For example, one alternative proposed is the recognition of revenue (sale) at the time of the formal contract of purchase, which is normally an executed purchase order. Another alternative suggested is the recognition of revenue as the production process is completed or some combination of both of these alternatives. At present none of these alternatives are acceptable. Revenue recognition is limited to exchange (sale). However, exceptions to this principle are acceptable in certain industries, which is discussed below.

INDUSTRY PRACTICES

Because of the nature of certain industries and because of a history of using practices peculiar to specific industries, exceptions to the principles above are permitted. However, the exceptions must be justified on the basis of the pecularities of the industries affected. The meat packing industry has traditionally valued its finished goods inventory at selling price. Thereby, revenue is recognized at the time the product is transferred to finished goods rather than at the time of exchange (sale). Although some meat packing firms are modifying this approach, it has been an historical practice in this industry and therefore has been accepted by the accounting profession.

The construction industry is another industry where exceptions are allowed. The exceptions must be tested and there must be adequate justification, but revenue may be recognized either under the full accrual method or the percentage of completion method. Although completion and exchange (sale) of a project may not take place for years, revenue may be recognized using the two methods noted because of the nature of the industry and the historical use of these methods. These exceptions must be justified and they must be consistently applied.[1]

THE AUDIT FUNCTION

Throughout this text references have been made to the reviewer of financial information. The reviewer in this context is the auditor. The auditor (reviewer) may be the outside certified public accountant (CPA) or the internal auditor. Although technically the two audit functions are separate, that is the internal auditors are employees of the firm and the outside auditors are independent contractors, in recent years there has been an increasing degree of cooperation between the two groups. The amount and extent of cooperation may vary from firm to firm, but the purpose is to share information and verify that the firm's financial information conforms to generally accepted accounting principles and is not misleading.

AUDIT INDEPENDENCE

One of the objectives of an audit of financial information is to provide the users of that information with an independent review. The independence of the auditor (reviewer) should prevent the information from reflecting the bias of the preparer. In other words, the information should be more objective than it would be otherwise.

In the construction industry there is always a need for outside funds. Lenders, for their own protection, ask that the construction firm provide them with a complete set of financial statements and that the financial information be reviewed by an independent CPA (outside the firm). If the firm has a large number of geographically

[1]Donald E. Kieso, and Jerry J. Weygandt, *Intermediate Accounting* (New York: John Wiley & Sons, 1977), pp. 21-38.

dispersed shareholders, an independent review of financial information is required by the federal securities acts. If the firm is privately owned but has a significant amount of debt, the creditors will require an independent review.

Although the internal auditors are employees of the firm, they usually report to management at high levels of the company and therefore are in a position to suggest modifications to financial information that, in their opinion, is biased. In medium to large construction firms the role of the internal auditor is a critical one. With the widely dispersed operations (projects) and the need to rely on field personnel, the judgment of the internal auditors is important to the presentation of unbiased financial information. Usually construction firms of this size have a highly professional internal audit staff (many of them holding the CPA certificate) in addition to employing an outside CPA firm to do the annual audit.

The accounting profession has defined the requirements for an outside CPA to consider himself (or herself) independent. The American Institute of Certified Public Accountants (AICPA) as part of its statement of the professional ethics of the CPA requires that the auditor (reviewer or outside CPA) be independent of the firm that he is auditing. This has been interpreted to mean that the outside CPA must not have any significant financial or personal interest in the firm that he (or she) has been engaged to audit.

The ethics of the accounting profession not only require that the auditor (reviewer) be independent, but there must not be an appearance of a lack of independence. What the auditor's ethics require is not only the fact of independence but also the appearance of independence. These two requirements are not identical. If for example, the auditor was in fact independent of the client and that fact could be verified by someone independent of the auditor, the ethical requirement is not met. If the auditor has been associated with the firm on some basis and this association gave the appearance of a lack of independence, the ethical requirement for independence has not been met. To the accounting profession the independence of the auditor (reviewer) is an important precept.

THE ROLE OF THE INTERNAL AUDITOR

The role of the internal auditor (reviewer) is less independent than that of the outside auditor. The internal auditor is a full time employee of the firm. Large firms in general and large construction firms whose operations are spread over a wide geographic area would be prudent to employ an internal auditor. It is impossible, under these conditions, for the firm's top managers to control all of the firm's operations. The internal auditors are the eyes and hands of the firm's top managers in these widespread operations.

Usually the internal auditor reports and is responsible to the firm's top managers. The reporting relationship can be single and direct or the relationship can be direct and indirect to a number of managers. Often the internal auditor reports to the vice-president of finance of the firm. However, he (or she) could report directly to the vice-president of finance and indirectly to a committee of the board of directors of the firm. Or the internal auditor could report directly to the vice-president of finance and report indirectly to the president and indirectly to a committee of the board of directors. Regardless of the specifics of the reporting relationships of the internal auditor, this person should report to the highest managers. This gives the auditor a degree of internal independence and the attention of those who are ultimately responsible for the publication of accurate financial information.

The internal auditor may do more than perform financial type audits. Management may ask the auditor to do special reviews and investigations, such as reviewing a control or a manufacturing process, procedure, or system. Internal auditors could be asked to evaluate the effectiveness of the purchasing function or some other function. Because they are available on a full time basis, management may ask them to participate in a number of situations. Although the internal auditor is primarily responsible for continuous management and financial audits of the firm, the only real limitations to his (or her) role are resources.

As the costs of external audits have increased, many firms have initiated programs of cooperation between the internal and external auditors. To the extent that the internal auditors can perform some of the audit steps, the fee of the external auditors is reduced. There is, however, a limit to the degree to which the external auditors can use the workpapers and the staff of the internal auditors. The degree of independence of the two groups of auditors is not the same. Because the internal auditors are employees of the firm the outside auditors must evaluate the services and information furnished by the internal audit staff. The responsibility for the certification of financial information remains with the outside auditors.

THE EXTERNAL AUDITOR

The purpose of the external audit is to provide the user of the financial information with an outside opinion relative to whether the information conforms to generally accepted accounting principles. Often, regardless of the legal status of the firm, an outside auditor's opinion is necessary if the firm is attempting to raise capital. In addition, firms whose stock is publicly held and traded freely across state lines must include an outside auditor's opinion with their published financial statements. Creditors have also changed their attitudes toward independent audits in the last decade. Even in privately held firms it is no longer unusual for the creditors to require, as part of the loan agreement, an annual audit of the financial statements by an outside auditor. The creditors regard the audit as an additional form of protecting their loan.

External audits of financial information cannot be regarded as inexpensive. Before an audit is begun, the outside auditor prepares an estimate or projection of the total cost of the audit. The projection is not binding. It is only an estimate. If questions arise or problems are encountered, the audit must meet professional standards regardless of a cost projection. The costs of an audit are based on the hours of audit time used times a billing rate plus expenses. The hourly billing rate is usually designated by type of auditor used, that is, there is a rate for a partner, a manager, a senior, a junior, and clerical personnel.

Along with other service costs the costs of audits have increased. If the firm to be audited has an internal audit department, the use of and cooperation between the internal and external auditors can reduce the costs of the outside audit. The internal auditors are paid regardless of what type of audit they are doing, while the cost of external auditors is incremental. However, the decision of whether to accept the work of other auditors is the prerogative of the engagement auditor. The ethics of the American Institute of Certified Public Accountants leaves the choice to the engagement auditor. Because the engagement auditor issues the audit certification, he (or she) has the responsibility of deciding whether to rely on other auditors or only his own staff.

The power and influence of the external auditor depends on the audit certification. The effect of the audit certification on the users of financial information is the auditor's most powerful tool. The reaction of the entities stock price and its creditors to the audit certification is what makes the auditor effective. The external auditor may be engaged only for compiling and reviewing the financial statements or for a full audit. The external auditor may issue an unqualified opinion (certification), a qualified opinion (certification), or a disclaimer of opinion.

If the engagement is a review engagement, the external auditor usually does not follow all the procedures required for a complete audit. However, the auditor must state the extent of his review so that the user of the financial information can determine how much reliance can be placed on the auditor's opinion. In the case of a compilation the auditor only classifies and summarizes the information furnished by the client. To issue an unqualified opinion (certification) the auditor must conduct an audit that meets the standards of the profession and the financial information must not materially depart from generally accepted accounting principles. The unqualified opinion of the external auditor is the most desirable opinion. The qualified opinion (certification) notes an exception or a qualification in the financial information. The qualification could be within the data themselves or it could be a limitation that was placed on the scope, testing, or the information provided to the auditor. In some cases the qualification of the auditor's opinion can adversely affect the market price of the firm's stock and have an unfavorable impact on the firm's loan agreements. A disclaimer of or adverse opinion by the outside auditor can have a disastrous effect on the firm's stock market prices and loans. The disclaimer of opinion indicates that the auditor was either prevented from performing an adequate audit or during the audit he (or she) noted a large number of material exceptions that prevent the financial information from conforming to general accepted accounting principles.[2]

In addition to the audit opinion, the external auditor issues an audit report and one or a series of management letters. The audit report is a detailed report of the auditor's findings. This report also includes the auditor's recommendations concerning adjustments to the financial records and changes in accounting policies or procedures.

The management letter is usually concerned with an evaluation of the firm's financial personnel and financial organization. Also, there may be more than one management letter. A different letter may be sent to different levels of management. Because of the nature of the management letter it is often not discussed with or made generally available to the client's personnel.

[2]Glenn A. Welsch, Charles T. Zlatkovich, and Walter T. Harrison, Jr., *Intermediate Accounting* (Homewood, Illinois: Richard D. Irwin, 1979), pp. 151-163, 856-858.

THE REGULATION OF THE ACCOUNTING PROFESSION

The accounting profession, like law and medicine, has been regulated from within. However, unlike in law and medicine, self-regulation has occurred with the approval of a federal (national) regulatory commission.

THE SECURITIES AND EXCHANGE COMMISSION

The only federal statutory regulator of the accounting profession is the Securities and Exchange Commission. The enabling legislation of the commission is very broad and establishes the responsibility to see that financial statements are "not misleading." Although the U.S. Congress was not as generous with its appropriation act, the SEC did not intend to enforce its charter or exercise its responsibilities. The SEC quickly delegated its power to the accounting profession and in effect said: "Accounting profession—regulate yourselves and if we do not like what you are doing we will let you know."

When, in the early 1970s, a commission was formed to study the effectiveness of the acccounting profession in the performance of its varied functions, the SEC was also incited to act. About 1971 the commission reported its findings, and from 1971 to the end of 1978 the SEC issued approximately 50% of its Accounting Series Releases which affirmed the SEC's position on various accounting principles and procedures. It is interesting to note that the SEC issued as many Accounting Series Releases in those 8 years as it did in the previous 37![3]

THE ROLE OF THE AMERICAN INSTITUTE OF CERTIFIED PUBLIC ACCOUNTANTS

The Securities and Exchange Commission transferred the responsibility for the development of the accounting profession and of accounting principles to the organization which represented the certified public accountants. At first the rules governing the treatment of accounting information were formulated through a committee of the AICPA. Eventually, however, an Accounting Principles Board (APB) was formed within the AICPA to review and recommend accounting principles. Although the APB's prime function was to recommend accounting principles to be followed, it remained an integral part of the AICPA.

As a result of the report of the Wheat Commission in 1971, the APB was replaced by a Financial Accounting Standards Board (FASB).

[3]Walter B. Meigs, A. N. Mosich, and E. John Larsen, *Modern Advanced Accounting* (New York: McGraw-Hill Book Company, 1979), pp. 702-719.

The members of the FASB are full time members and are paid accordingly. The FASB is independent of the AICPA and has full responsibility for determining accounting principles, procedures, and reporting requirements.

In addition, each state has a State Board of Accountancy. The state boards determine the standards for becoming a certified public accountant and administer the uniform national certified public accountancy examination. The state boards are also responsible for defining the ethical requirements for the certified public accountants within the state. The state boards work closely with the AICPA.

Usually the income tax effect of accounting policies is separated from the treatment of accounting and financial information in financial statements. But the Internal Revenue Service (IRS) has influenced the accounting treatment of various financial information. The IRS has affected the accounting treatment of such items as inventory valuation, particularly the use and application of the last-in-first-out (LIFO) method of inventory valuation. The IRS has also influenced depreciation policies and the treatment of tax credits.

Often students and those unfamiliar with the practice of accounting think that the certified public accountant can require managers and preparers of financial information to conform to the requirements of generally accepted accounting principles. However, it must be recalled that the effectiveness of the reviewer (auditor) depends on the ability of his report to influence investors, shareholders, and creditors. The implementation of the results of the accountant's work depends on the effect he (or she) may have on the client through the effect of the audit certification on stock market prices and loan agreements (debt agreements).

THE CHOICE OF ACCOUNTING METHOD FOR THE CONSTRUCTION INDUSTRY

The construction industry, like other industries, has a number of choices of bases for preparing financial information and income tax returns. Firms in the construction industry have the choice of using one method for income tax purposes and another method for preparation of financial information.

FOR FINANCIAL STATEMENT PURPOSES

The cash basis is widely used by small construction firms. Under this method, transactions are recorded when cash is exchanged.

That is, the entries in the financial records are not made until bills are paid or until cash is received from customers. For small firms the cash basis provides adequate information for both income tax preparation and for the firm's owner and/or manager.

For larger firms, firms that are classified as medium or smaller, the modified cash basis is sufficient. With the modified cash basis, transactions are recorded when cash is exchanged except for equipment, inventory, accounts receivable, purchases and sales. Equipment is depreciated and capitalized (put in the balance sheet) as an asset. Sales made on account are recorded in the accounting records with the related receivable. The recording of the credit sale is necessary for the proper establishment of inventory. If the firm has a substantial inventory of supplies and construction material that does not belong to the general contractor or owner (customer), purchases on account must also be recorded. The recording of on account purchases is necessary to properly establish inventory values. The modified cash basis usually provides enough information for both income tax purposes and for the firm's owner/manager.

The completed contract method of accounting is allowable for use by medium and large construction companies. Under the completed contract method the revenue, expenses, and net income from a project are not recognized until the project is complete. In the case of a structure such as a building, completion usually means beneficial occupancy by the owner (customer). If progress payments are available to the contractor, he (or she) may treat the payments as revenue to the extent of the applicable cost incurred. The owner (customer) will usually allow progress payments to the degree that the payments reimburse the contractor for the actual costs incurred, and less the agreed upon retention withheld, the payments do not decrease the contractor's desire to complete the project.

The completed contract method is not recommended for financial statement purposes unless there are enough large projects in progress to result in the preparation of meaningful financial statements. If, for example, at least one project will be completed each fiscal year from which revenue, expenses, and a reasonable net income can be reported, the completed contract method is useful as a method for the preparation of financial statements. If, on the other hand, the completed contract method would result in reporting no revenue, expenses, or net income for a fiscal period, it is not recommended. The reporting of zero net income requires that the progress payments received, if any, be included in the statements as unearned income (an advance from the customer/owner) and that all costs incurred be treated as construction in progress (inventory or receivable).

The percentage of completion method of accounting for financial statement preparation is the method usually recommended. Under this method revenue is accrued (the contract amount for the construction project) on the basis of the engineering estimate of the percent of completion of the project. The actual cost is assigned to the period based on that incurred to the statement date. The difference is the net income or loss on the project through the reporting date. With this method the financial statements reflect the construction firm's performance for the reporting period. The revenue and cost information is recorded on a cumulative basis, with the reporting period reflecting the difference between the last report and the cumulatives through the current period. Any progress payments can be recorded as either revenue earned or as unearned revenue (an advance). If the progress billings and payments are treated as earned, the accrual of revenue would be the difference between the progress payments received and the percent of contract completion. If the progress billings and payments are treated as unearned, they would be carried as a liability account titled unearned revenue until the project is complete. At that time the accrued revenue would be offset against the unearned revenue.

The percentage of completion method is recommended for financial statement purposes because it provides for some meaningful financial statements during the construction period. This is particularly important if the ownership and management of the firm are separated or if the construction firm is seeking outside financing in the form of loans. It is also a useful method for managing a construction firm whether the manager is the owner or not. Percentage of completion financial statements are useful to report management stewardship during the construction period. Although the project may not be completed, the percentage of completion method results in useful progress reports. The completed contract method does not report performance until the project is complete. Through the percentage of completion method the owners are informed of the progress during construction.

Also, if the construction firm has or is seeking a loan, the lendor will want to be kept informed of the firm's progress. The lendor will not want to wait until the project's completion to learn of its success or failure. The percentage of completion method best serves this need. In addition, the managers will want to know if the firm and its projects are proceeding satisfactorily as soon as possible. The managers will not want to wait until the project or projects are complete. The percentage of completion method is less easily manipulated because outside and independent engineers can be hired to verify measures of physical completion.

The full accrual method is available for use for financial statement purposes under generally accepted accounting principles. However, this method is not recommended except when the construction firm sells a product, such as concrete or forms. The full accrual method is usually based on product shipment, time, or cost incurred at the statement date. The full accrual method has its usefulness, but is not recommended for project type construction firms, because it can result in misleading financial statements.

FOR INCOME TAX PURPOSES

The cash basis is useful for income tax purposes when it will result in the lowest tax liability and when it is acceptable for tax preparation. Usually the cash basis serves well if the net cash inflow is relatively stable and the individual and corporate tax rates remain relatively constant. Under some conditions, with a constant corporate tax rate, delaying the tax as much as possible may be more profitable than using the cash basis. That decision is based on a cost-benefit analysis for the firm and the way the firm is organized. However, one of the strongest arguments for the use of the cash basis of accounting for income tax preparation is that the income tax liability is recognized when the cash is available to meet the liability (assuming profitable operations and sound cash management).

The modified cash basis is often required as a method of accounting for income tax preparation. If the construction firm maintains any significant amount of inventories, it is necessary to accrue shipments made on account and to accrue the purchase of material on account. These accruals are needed to properly establish an inventory value. If equipment is purchased, it is necessary to capitalize the equipment on the balance sheet and recover the cost through depreciation charges to operations and projects. The modified cash basis has the same advantages and disadvantages as the cash basis. The main difference is the need to establish inventory values which causes the need for the use of the accrual method of accounting for some accounts and for the preparation of the income tax returns. The use of the modified cash basis of accounting, especially in the early stages of a project, can cause the need for cash to make income tax payments before the project billings are sufficient to generate the cash.

The completed contract method is the best method to elect for income tax purposes under most circumstances. The completed contract method delays the income tax liability until the project is completed. In addition, at the completion of the project, assuming that the project is completed according to the specifications in the contract, the customer pays the retention as well as the final billing. Therefore, the cash is available to pay the income tax liability. If the construction firm is organized in the corporate form the completed contract method is more acceptable because corporate tax rates are not graduated. However, the firm's managers must estimate what the corporate income tax rates will be at the completion of the contract to select the most profitable accounting method for the firm. If the construction firm is organized as a proprietorship or a partnership, more analyses are needed to determine whether the completed contract method is the most appropriate. Because individual income tax rates are graduated, the individual tax rates at contract completion must be estimated, the individual rate status of each owner must be projected at contract completion and these must be weighed against the time value of money to the firm from the delay in the payment of income taxes.

The percentage of completion method is usually not recommended for income tax preparation. This method entails the accrual of project revenue and therefore some income tax liability during the construction period (assuming a net income). It is useful as an accounting method for income taxes in some cases, if the construction firm is organized as a proprietorship or a partnership because less tax liability can result if the owner's income tax bracket is less than with the completed contract method. However, even with progress billings and payments the cash may not be available to pay the income tax liability because of retention.

The full accrual method is not recommended for income tax preparation purposes. This method usually results in a higher tax liability during the construction period than the percentage of completion method and therefore is less desirable. Again, if progress billings are allowed by the contract and progress payments are received on a timely basis, because of the retention provisions cash may not be available to pay income tax liabilities.

SUMMARY

In this chapter generally accepted accounting principles are reviewed. These are the principles that underlie the preparation of all

financial statements. The role of the auditors in verifying compli-
ance with generally accepted accounting principles is described. The
need for auditor independence and the role of internal and external
reviewer (auditor) are discussed. The agencies and organizations
that review and amend accounting requirements are listed and their
functions and relationships are examined. In addition, in this
chapter the methods of accounting for financial statement purposes
and for preparation of income tax returns available to construction
firms are analyzed. Included in the analysis is a recommendation of
which methods, under given circumstance, are of most benefit to the
construction firm for the preparation of financial statements and
income tax returns.

QUESTIONS

1. Describe and discuss the accounting principle of matching.
2. What is the purpose of the accounting requirement for full
   disclosure?
3. Discuss the accounting principle of materiality and describe its
   relationship to the other accounting principles.
4. Define the accounting principle of conservatism.
5. How does the accounting assumption of the going concern
   affect accounting thought and the evaluation of the firm's
   financial statements?
6. What is the historical cost assumption?
7. Discuss and describe, in detail, the accounting principle of
   revenue realization.
8. What is the role of the internal auditor?
9. What is the role of the external auditor (reviewer)?
10. Describe the various types of audit certifications.
11. Describe the role of the Securities and Exchange Commission
    in regulating the accounting profession.
12. What is the history of the determination and codification of
    accounting principles?
13. Discuss the role of the AICPA in determining accounting
    principles.
14. What is the function of the FASB?
15. Discuss the alternative accounting methods available to a firm
    in the construction industry for the preparation of financial
    statements.

16. Discuss the alternative accounting methods available to a firm in the construction industry for the preparation of income tax returns.

17. Which accounting methods for the preparation of financial statements and for the preparation of income tax returns are recommended and why?

REFERENCES

Anthony, Robert N., and James S. Reece, *Management Accounting* (Homewood, Illinois: Richard D. Irwin, 1975).

Chorba, George J., *Accounting For Managers* (New York: American Management Association Extension Institute, 1978).

Gordon, Myron J., and Gordon Shillinglaw, *Accounting: A Management Approach* (Homewood, Illinois: Richard D. Irwin, 1974).

Kieso, Donald E., and Jerry J. Weygandt, *Intermediate Accounting* (New York: John Wiley & Sons, 1977).

Meigs, Walter B., A. N. Mosich, and E. John Larsen, *Modern Advanced Accounting* (New York: McGraw-Hill Book Company, 1979).

Montgomery, A. Thompson, *Managerial Accounting Information* (Menlo Park, California: Addison-Wesley Publishing Company, 1979).

Welsch, Glenn A., Charles T. Zlatkovich, and Walter T. Harrison, Jr., *Intermediate Accounting* (Homewood, Illinois: Richard D. Irwin, 1979).

CHAPTER 4

PURCHASING, ACCOUNTS PAYABLE, AND CASH DISBURSEMENTS

Purchasing, accounts payable, and cash disbursements are all related. The first step in the process is the need identification and placing of the purchase order. Then, when the good or service is received, a liability is recognized by the purchaser. After the liability has been properly established, a cash disbursement, for the proper amount, is made to the supplier or vendor.

PURCHASING

It is not necessarily the purchasing department's function to recognize the need for a product or service. However, once a need has been established, it is the responsibility of the purchasing department to acquire the good or service in the quantities requested, at the lowest possible price within the required specifications. The purchasing function is responsible for knowing the market, that is, the suppliers of goods and services and the prices.

THE PURCHASE OF MATERIAL AND SERVICES FOR GENERAL USE

The purchase of material and services for general use is usually the function of the purchasing department in the construction industry. Long term agreements may be executed by purchasing with vendors to provide repair services for machinery, equipment, vehicles, and office machines. For material such as forms, small tools, fasteners, and other items not peculiar to a particular construction project, purchasing will buy for inventory and the material will be withdrawn from inventory for project use. With these types of purchases a centralized purchasing department is adequate to meet the firm's needs, although a method must be provided for the purchase of emergency needs at the project site by field personnel.

Because of the costs, a project cannot be delayed and penalties incurred or excess costs paid when a vendor sends the wrong material or wrong quantity to the project site. Some decentralized purchasing is necessary to allow the field superintendent to meet his responsibilities. This can be accomplished by attaching a decentralized purchasing agent to the larger projects or through the use of field purchase orders issued by the project manager. Copies of the field purchase order and evidence of receipt of the material must be returned to the central purchasing department and to the accounting department. An example of a field purchase order is included as Exhibit 4.

The purchase of general use equipment is more complex. The centralized purchasing department can make the necessary vendor contacts; however, the evaluation of the lease or buy or the type of equipment to purchase within the technical specifications requires the participation of the financial department. At this point questions of cost, utilization, and return are answered. When these questions are resolved the purchasing department can conclude the transaction with the vendor. Field personnel usually participate in this process to verify that the equipment purchased meets their needs.[1]

THE PURCHASE OF MATERIAL SPECIFICALLY FOR CONSTRUCTION PROJECTS

These materials, including equipment, are not of a general use nature and are needed for one or a few construction projects. In some cases, these materials can be purchased by centralized purchasing but often it is necessary to decentralize purchasing and have the purchase made at the project site. This is necessary because of the conflict that often occurs between design and construction.

Purchasing at the job site can be organized in a number of ways. The purchasing department or part of it may be decentralized and be located at the job site or located convenient to a number of job sites.

[1]Charles T. Horngren, *Cost Accounting: A Managerial Emphasis* (Englewood Cliffs, New Jersey: Prentice-Hall, 1977), pp. 464-477.

Or the purchasing may be done by field personnel. In this case, the field personnel can choose the most suitable vendors or they may be required to purchase from a list of approved vendors. When field personnel purchase material a field purchase order is usually required. The field purchase order is the contract between the construction firm and the vendor.

Field purchasing causes problems of control for the financial department. When field personnel perform the purchasing function, the roles of buyer and user are merged. The merging of these roles can lead to compromise of the firm's assets and in some cases to bankruptcy through fraud. The temptation placed before field personnel is almost irresistible. Although the use of decentralized purchasing overcomes some of these temptations, if the decentralized purchasing agent is responsible to the field supervisor, through collusion, the effect can be the same as if purchasing were not decentralized. If the decentralized purchasing department reports to centralized purchasing, the risk of compromise is less but the possibility of off specification purchases and project delays is increased. The final organization structure must meet the needs of each construction firm.

SUBCONTRACTING

Subcontracting is extensively used in the construction industry. The purchase of subcontracting is unlike the purchase of material or of services. Subcontracting is usually the purchase of services and material. In some construction firms all of the work is done by subcontractors and the construction firm manages the project and supervises and coordinates the work of the subcontractors. Subcontractors are usually proficient in certain areas of construction and work for a number of general contractors. The use of subcontractors is more economical because the construction firm does not need to invest in the knowledge, personnel, and equipment used for only a portion of the construction project.

The initiation of subcontracts from central purchasing is not universally feasible. At a minimum field personnel must be continuously consulted. It is extremely risky to award subcontracts solely on price. It is a mistake to choose subcontractors who could delay construction project completions. If the subcontracting services required are repetitive, with feedback from the field about subcontractor performance, central purchasing can choose and award the subcontracts. Another alternative is for central purchasing to furnish field personnel with a list of approved or qualified subcontractors from which to select.

If the type of subcontracting required is nonrepetitive, field personnel must have more participation in the selection process. The timely completion of the construction project is the responsibility of the field personnel and a poorly selected subcontractor can cause a completion date to be missed as well as incur the expense of rework for below specification performance. When choosing the proper procedures for the selection of subcontractors, it is necessary to balance the responsibility of the field personnel for timely completion of the project within specifications against the risk of collusion between field personnel and subcontractors.

ACCOUNTS PAYABLE

It is the responsibility of the accounts payable personnel to determine, as accurately as possible, that the firm does not recognize invalid obligations. It is also the responsibility of this group to record, in the firm's financial statements, all legitimate accounts or debts payable to vendors, suppliers, or subcontractors.

THE NEED FOR PURCHASE ORDERS

Purchase orders are necessary to support the recognition of a payable that will be eliminated through the expenditure of the firm's cash. Material amounts of cash should be disbursed for only recognized liabilities. No material liability should be recognized without a purchase order. Whether the purchase order is issued by central purchasing or by field personnel, it is necessary to support the expenditure of funds. Without the purchase order the conditions of purchase are not available to those who disburse funds. The purchase order is needed to determine the type of materials ordered, the quantities ordered, the price agreed upon, the shipment terms, the date of delivery, and the terms of payment. A copy of the purchase order (which is the contract between the firm and its supplier) must be furnished to the financial section of the construction firm for use by the accounts payable section. Both the receiving report and the vendor's invoice will be matched against the purchase order by accounts payable. In the construction industry, because of the wide geographic locations of projects, this requirement is often overlooked. At the least the lack of purchase orders can lead to delay in payments and to poor vendor relations with, perhaps, the withholding of critical material by the vendor. The use of only an authorized signature on the vendor's invoice as purchase authorization can lead to compromise of the construction firm's assets. For significant purchases, the availability of modern communication devices causes the lack of purchase orders because of wide geographic dispersion to be questioned. An example of a purchase order is given as Exhibit 5.

THE RECEIVING FUNCTION

The receiving function is more complicated in the construction industry. For general use materials a central receiving department can be used, with the material reshipped to the various using locations or drop shipped at the using location with a report on receipt sent to central receiving. Although inspection is necessary, for general purpose materials the knowledge is usually available at the central receiving as well as other locations.

Regardless of where the material is shipped, a record of receipt or of performance of the service (i.e., subcontracting) is necessary for payment. A construction firm should not disburse funds to suppliers and subcontractors without a record of receipt or performance. When special materials, used only for a single project or a few projects, are involved the problem of receipt is further complicated within the construction firm. Central receiving may not have the knowledge needed to properly inspect special purpose materials and, in addition, if special purpose materials are received centrally, the central receiving may not know where to ship them for use.

Usually, special purpose materials are shipped directly to the project or projects for which they were purchased. This requires heavy reliance on field personnel to inspect the material for conformance to specifications and for receipt of the quantities ordered. Reliance on field personnel for this information could result in a breach in the firm's internal control system. However, in these situations, field personnel forward proof or approval of receipts or performance to the finance department for payment of the vendor's invoice.

In construction firms, because of the numerous and diverse locations, some reliance on field personnel is necessary. It is prudent to evaluate the use of a central receiving location and the costs of reshipment against the risk of delaying the completion of a project and its accompanying cost with the possible compromise of the firm's resources. There is a possibility that the records to support payment that are found in a manufacturing and retailing environment are not available in a construction environment. There is a tendency to rely on field personnel in construction; however, it must be remembered that the prime responsibility of field personnel is to complete the project within the contract terms and completion date.

PREPARATION OF THE VOUCHER PAYABLE

The timely preparation of the voucher for vendor payments depends on the receipt and processing of the supplier's invoice. Within firms outside the construction industry, the purchase order, the receiving slip, and the vendor's invoice are matched before the voucher is prepared. This matching process verifies that the materials ordered were the materials received. Matching also verifies that the quantities ordered were received and that the firm was invoiced for the proper quantities at the price ordered.

In the construction industry all of the documentation may not be available. In that case, invoices should be separated by amounts and the larger invoices given special attention. The invoices can also be separated into those for special purpose materials and those for general purpose material. The invoices for smaller amounts may be processed for payment (the voucher prepared) with only one approval. Invoices for larger amounts should have the approval of field personnel and of a field superintendent or manager before the voucher for payment is prepared. When either the documentation has been satisfactorily matched or the approvals has been secured, the voucher can be prepared by the financial or accounting department.

The vouchers payable are sequentially numbered to prevent payment twice and are perforated after payment. The voucher cover contains all the necessary information for payment, such as the vendor's name, the amount payable, the accounts to be affected, and approval by a member of the management of the accounting and finance department to verify that the voucher has been properly prepared. This process is also applicable to processing invoices for services and subcontracting.[2]

It is the responsibility of the accounting and finance department to take advantage of all discounts offered by suppliers. These are usually cash discounts for timely payment of the invoice and are not offered by subcontractors. The suppliers will specify the terms of the cash discount, such as allowing a 1 or 2% cash discount on the total invoice amount if paid within 10 days of the invoice date. The purchase discount can be accounted for in one of three ways.

One method is to record the purchase amounts gross and to recognize the discount when payment is made within the discount terms. For example, if it is assumed that a purchase of $1,000 is made

[2]William W. Pyle, John Arch White, and Kermit D. Larson, *Fundamental Accounting Principles* (Homewood, Illinois: Richard D. Irwin, 1978), pp. 226-247.

with terms of 1% if paid within 10 days of the invoice date, the entries would be as follows. At purchase:

	Debit	Credit
Purchases		$1,000.
Vouchers payable		$1,000.

At time of payment:

	Debit	Credit
Vouchers payable	$1,000.	
Purchase discounts income		$10.
Cash		$990.

If the net method is used for entries are as follows. At purchase:

	Debit	Credit
Purchases	$990.	
Vouchers payable		$990.

At time of payment within the discount terms:

	Debit	Credit
Vouchers payable	$990.	
Cash		$990.

At the time of payment after the discount period has expired:

	Debit	Credit
Vouchers payable	$990.	
Discounts lost	$ 10.	
Cash		$1,000.

If the allowance method is used the entries are as follows. At purchase:

	Debit	Credit
Purchases	$990.	
Allowance for purchase discount lost	$ 10.	
Vouchers payable		$1,000.

At the time of payment within the discount period:

	Debit	Credit
Vouchers payable	$1,000.	
Allowance for purchase discounts lost		$10.
Cash		$990.

At the time of payment after the discount period has expired:[3]

	Debit	Credit
Vouchers payable	$1,000.	
Cash		$1,000.

Some firms have a policy of taking all cash discounts regardless of when the vendor's invoice is paid. However, with this policy as with all other policies regarding payables, the effect on vendor relations must be considered. Any policy that alienates vendors may be detrimental to the firm.

GENERAL CASH DISBURSEMENTS

One method of internal control over cash is the use of a limited number of payment methods and sources. A widely used method is the voucher payable. Under these circumstances a payment cannot be made without the preparation of a voucher payable (except for petty cash, which is discussed below). The use of the voucher strictly limits the sources and number of payment methods. For construction firms the problems of the use of a voucher system are compounded by the number and diverse locations of construction projects. The best approach would be field preparation of voucher information and a centralized cash disbursement fund.

If that is not possible, there can be field disbursement funds that cannot be disbursed without the preparation of a voucher at the project location. A periodic audit can enforce the requirement for proper documentation at the project level. In addition, petty cash funds can be located where needed to provide for field disbursements. As the number of disbursement accounts increases and the number of petty cash accounts proliferates, better controls are required. The risk of loss from compromise must be balanced against the costs of controls and the most optimum mix for the firm selected.

Despite the use of a voucher system, the use and issuance of checks must be strictly enforced. The unissued checks must be kept in a locked location with access limited to one or a few individuals. Each check number must be accounted for. Timely bank reconciliations are a necessary part of internal control of cash. The prompt reconciliation of the bank statement is a deterrent to the unauthorized use of checks. The reconciliation can be done manually (usually

[3]Glenn A. Welsch, Charles T. Zlatkovich, and Walter T. Harrison, Jr., *Intermediate Accounting* (Homewood, Illinois: Richard D. Irwin, 1979), p. 358.

by someone other than the cash disbursement section employees) or with data processing equipment.

SIGNATURES AND BANK ACCOUNTS

Construction as well as other firms have one or a number of general fund bank accounts. If disbursements can be made from one central location, only one general fund is required. The general fund is the source of disbursements for all purposes. Additionally, special bank accounts are used for payrolls and subcontractors. These special accounts are funded through the general disbursement fund for the amount of the payroll or the payment to subcontractors. Through this technique these bank accounts should equal zero or clear after all checks issued have been cashed. (Most states have rules concerning uncashed and unclaimed payroll checks.) The use of these special bank accounts removes some of the burden of reconciling the general fund balance and, by isolating the totals of each, provides more control over disbursements for payroll and subcontractors (assuming that payments to subcontractors are sufficiently large to warrant a separate bank account).

In some cases, the checks are printed and signatures applied on data processing equipment as the disbursement journals are being created. Limiting the access to the signature plates in these instances and witnessing the use of the plates are mandatory. If the checks are printed and then processed in a separate check signing machine, the same controls are necessary. Whether checks are signed on data processing equipment or on a separate check signing machine there should be a dollar limit printed on the check limiting the amount for which a check can be written through these devices. Checks for amounts larger than these should require one or two manual signatures as needed for internal control of cash. Checks signed manually should be printed on a different color and/or check forms than any others. Manual check authorization signatures usually vary for the amount of the check and the management level of the signer.

FLOAT

Particularly in an environment of high interest rates and cash shortages, the construction firm must make a maximum use of float. Float is the time between the issuance of a check and the deduction of the check from the firm's bank accounts. The objective is to prolong the time it takes a check to clear at the firm's bank. One method to increase clearance time is to use an account at a remote bank which is in a different federal reserve district from the bank in which the check will be deposited. Another method is to send the check without the proper zip code and thereby increase transit time or to send the check to the payee's home office rather than to the designated address or to the designated lock box.

On the other side of the issue, if the firm must reduce the float of payments from customers, the customer can be directed to make payment to a lock box. A lock box is a mailing address controlled by a bank that is close to the firm's customers. The bank will deposit the checks to the firm's account and send copies of the checks to the firm for use in cash application (applying the payments to the customers account). This results in quick check clearance, making the funds available to the firm. The fastest method of fund transfer from customers is to use the wire transfer services of the depository banks. Payment can be made and the funds become available to the firm in a matter of hours or a day. The difficulty is that most of the firm's customers will be reluctant to reduce their float and pay by wire transfer without some incentive and this technique is expensive and unsuited to the transfer of small amounts.

IMPREST FUNDS (PETTY CASH)

Imprest funds are used to avoid the need to issue a large number of checks for small amounts of money. Instead, the cash in the imprest fund is used to buy small items and to pay for small invoices. The fund is also used to pay for small delivery charges, such as parcel post. Funds can be located at each construction project, if necessary, and an on site employee can be made responsible for the fund. Once the size of a fund has been determined, that amount remains in the general ledger until the fund size is changed. The reimbursement of the fund, for cash expended, is charged to the various expense accounts for which the money was used. In this way the fund can be verified at any time by checking the amount of the fund in the general ledger and asking the petty cashier (or responsible employee) to produce the cash or receipts up to the established amount of the fund.

The journal entries required are as follows:

	Debit	Credit
Petty cash	$500	
Cash		$500
To establish the fund.		

Supplies expense	$100
Delivery expense	100
Miscellaneous expense	100
Cash	$300

To reimburse the fund for cash expended

Procedures for control of the imprest funds must be established and enforced to prevent compromise of the cash. Imprest fund controls should include at least the following:

(a) Establish the fund for the minimum amount needed to meet the demands for small payments. Keeping the imprest fund small reduces the temptation to employees and the opportunity for compromise.

(b) Access to the imprest fund must be limited to one or at most two responsible employees. And a record must be available on when each employee has responsibility for the fund. Establishing exact responsibility reduces the possibility of misuse of the fund.

(c) Receipts and approvals must be obtained for any expenditures from the fund and a transfer receipt should be prepared when possession of a fund is transferred from one employee to another.

(d) The cashing of personal or payroll checks from the imprest fund must be prohibited. The cashing of checks requires a fund amount that is larger than necessary, thereby increasing employee temptation to compromise the fund and limiting the amount of cash available for other purposes. In addition, insufficient fund checks may be cashed through imprest funds.

(e) A manager or supervisor should be assigned the responsibility for making surprise counts of the fund. The manager should verify that the cash plus any receipts equal the size or amount established for the fund.

(f) The imprest fund must be audited at least once a year by both the internal and external auditors. The auditors must count the cash, check all receipts, and verify that the required signatures were obtained.[4]

[4]Pyle, *op. cit.*, pp. 180-184, 231-234.

SUMMARY

This chapter is a discussion of the purchase of materials and services, the recognition of the liability for those purchases, and the payment to the vendor or supplier. The emphasis is on the control of the materials and services purchased to assure the construction firm's owners and managers that funds are disbursed for only authorized purposes. The discussion includes the use of centralized versus decentralized purchasing, the need for a voucher system to control payables, and the use of general cash disbursements, special cash disbursements, and petty cash funds. Methods for control at each step in the disbursement process are suggested and the necessary journal entries are given.

EXHIBIT 4.   An Example of a Field Purchase Order

Name of Firm Issuing the Purchase Order

Name and address of vendor.   PO Date _____

                                               PO # _____

Promised Delivery Date	Method of Shipment	Shipment Terms	Payment Terms	Sales Tx Ex #

Quantity Ordered	Description	Material Code	Unit Price	Total Cost

Signature   Signature

Notes:
(1) The signatures are those of field personnel.

(2) Copies:
     a. Vendor. b. Accounting. c. Purchasing. d. Field Superintendent e. Employee preparing order.

EXHIBIT 5. An Example of a Purchase Order

Name of Firm Issuing the Purchase Order

Name and address of vendor.

Vendor Code # _____

Purchase Order Date _____

Purchase Order # _____

Buyer Code # _____

Promised Delivery Date	Method of Shipment	Shipment Terms	Payment Terms	Sales Tx Ex #

Quantity Ordered	Description of Material or Service	Material Code	Unit Price	Total Cost

_____   _____

Approval Signature   Buyer's Signature

Notes:

(1) The order may be prepared for one type of material or service or it may be prepared for all materials and services from a single vendor.

(2) Copies:
   a. Vendor or Supplier. b. Purchasing File. c. Receiving Department. d. Ordering Department. e. Accounting Department. f. Buyer.

QUESTIONS

1. What are the differences between centralized and decentralized purchasing?

2. Why is a method for field purchases required in the construction industry?

3. What are the differences between subcontracting and other purchases?

4. What makes subcontracting so important to the construction industry?

5. What is the role of purchasing in the cash disbursement process?

6. What are accounts payable and what is their role in the disbursement process?

7. What are vouchers payable?

8. What documents are required for a vouchers payable system?

9. Does the receiving function have a role in the cash disbursement system? If so, what is its role?

10. Describe a typical voucher in a vouchers payable system.

11. Why is cash disbursement control more difficult in the construction industry?

12. How many and what types of bank accounts are used for cash disbursements?

13. What is the purpose of a bank reconciliation?

14. Discuss the signature controls needed in a cash disbursement system.

15. What is the meaning of the term float and how is it used?

16. What is the purpose of a petty cash fund?

17. How does a petty cash fund work?

18. To prevent compromise of the imprest fund, what controls should be used?

REFERENCES

Anthony, Robert N., and James S. Reece, *Management Accounting* (Homewood, Illinois: Richard D. Irwin, 1975).

Chorba, George J., *Accounting for Managers* (New York: American Management Association Extension Institute, 1978).

Gordon, Myron J., and Gordon Shillinglaw, *Accounting: A Management Approach* (Homewood, Illinois: Richard D. Irwin, 1974).

Horngren, Charles T., *Cost Accounting: A Managerial Emphasis* (Englewood Cliffs, New Jersey: Prentice-Hall, 1977).

Kieso, Donald E., and Jerry J. Weygandt, *Intermediate Accounting* (New York: John Wiley & Sons, 1977).

Meigs, Walter B., A. N. Mosich, and E. John Larsen, *Modern Advanced Accounting* (New York: McGraw-Hill Book Company, 1979).

Montgomery, A. Thompson, *Managerial Accounting Information* (Menlo Park, California: Addison-Wesley Publishing Company, 1979).

Pyle, William W., John Arch White, and Kermit D. Larson, *Fundamental Accounting Principles* (Homewood, Illinois: Richard D. Irwin, 1978).

Welsch, Glenn A., Charles T. Zlatkovich, and Walter T. Harrison, Jr., *Intermediate Accounting* (Homewood, Illinois: Richard D. Irwin, 1979).

CHAPTER 5

THE FINANCIAL STATEMENTS AND FISCAL PERIOD

There are four basic financial statements. One is the statement of financial position, another is the income statement, a third is the statement of retained earnings, and the last is the statement of changes in financial position. The statement of financial position lists the firm's assets, liabilities, and owner's equity at a certain point in time and is the oldest of the statements. The income statement summarizes the firm's performance for an accounting period. Included in the income statement are sales, cost of sales, expenses, income taxes, and net income. For the construction firm the income statement usually includes project revenue, direct project costs, allocated project costs, general or central expenses, income taxes, and net income.

The statement of changes in financial position is an analysis of the sources and uses of funds by the firm for the accounting period. This statement may be prepared on a cash or the more widely used working capital basis. All four of the financial statements are prepared, at a minimum, at the end of the fiscal year. However, many firms present and prepare financial statements at the end of the fiscal month and quarter as well as for the fiscal year. The selection of the fiscal period can and does influence the interpretation of the financial statements because of the effect of that selection on content and comparability from period to period.

THE STATEMENT OF FINANCIAL POSITION (BALANCE SHEET)

The statement of financial position is prepared as of a certain date, that is, the amounts listed in the statement are as of the statement date. The balances are a picture of the firm at that point in time. However, from that date forward the firm's status is continuously changing. Therefore, comparative statements are more meaningful for evaluating the firm. The name of the statement of financial position is relatively new; this statement was formerly titled the balance sheet.

THE STATEMENT OF FINANCIAL POSITION CLASSIFICATIONS

All accounts that are included in the statement of financial position are real accounts. That is, these accounts are not closed at the end of the accounting period but remain open in the general ledger until they are no longer used. The type of accounts that are included in the statement are asset, liability, and owner's equity accounts. These classifications are further subdivided into current and noncurrent, except for owner's equity.

The asset accounts are divided into current, long term investments, operational assets—tangible and intangible, and other assets. Examples of current assets are cash, marketable securities, notes receivable, accounts receivable, and inventory. Long term investments usually consist of bond sinking funds, investments in subsidiaries, and operational assets held for future use or not currently used in operations. Tangible operational assets include land, buildings, machinery, and equipment used to operate the business. Examples of intangible assets are patents, copyrights, trademarks, and goodwill. Other assets are such accounts as prepaid insurance and assets that do not fit in any other classification.

The liabilities are divided into current and long term. Current liabilities include notes payable, trade payables, other payables, and the current portion of long term debt. Examples of long term liabilities are long term notes payable, bonds payable, mortgages payable, liability under product warranties, and minority interest in subsidiaries.

The owner's equity accounts are listed in the balance sheet by source. The source of owner's equity is contributed capital, earned capital, and miscellaneous sources. Contributed capital includes the par value of the common and preferred stock sold and the paid in capital in excess of par value from the sale. Retained earnings are listed next, and this is the account in which the earnings from the business remain until paid to the owners. An example of a miscellaneous account in the owner's equity is the purchase of the firm's own stock. If the stock purchase is accounted for by the cost method, the last item in the owner's equity section is a reduction in the total owner's equity for the cost of the firm's stock purchased.

The statement of financial position for the construction firm may differ slightly from that of the nonconstruction firm. Most of the accounts found on the statement of financial position for other firms appear on a statement for a construction firm. The major areas of difference are in the inventories and the accounts receivable. For a construction firm the work in progress is very large and the finished goods very or relatively small. If the construction firm uses the completed contract method, the work in progress inventory is extremely large. The work in progress inventory consists of all the construction projects in process (uncompleted) and there is a detailed subsidiary ledger by construction project.

The accounts receivable also is detailed by construction project and is separated into at least four categories: (a) accounts receivable—construction projects not yet invoiced to customers (general contractors or owners); (b) accounts receivable—projects with progress billings submitted to customers; (c) accounts receivable—projects with final invoices submitted to customers; and (d) accounts receivable—retention withheld by customers. In the case of the construction firm, the accounts receivable can be very large depending on the size of the construction projects.

The statement of financial position is normally prepared in one of two forms: the account form or the report form. The account form is prepared with the assets on the left side and the liabilities and owner's equity on the right side of the statement. The report form is prepared vertically. The assets are listed first, the liabilities are listed under the assets, and the owner's equity is listed under the liabilities. Regardless of which form is used for the statement, the reader must recall that the total of the assets must equal the total of the liabilities plus the total of the owner's equity.

THE USES OF THE STATEMENT OF FINANCIAL POSITION

The statement of financial position has many uses and many users. Creditors, for example, are interested in the number and value of the firm's assets and in the amount of current assets versus the amount of long term debt to owner's equity. They are interested in confirming that the firm does not have more outstanding debt than the owner's investment can support. Creditors also want to verify that the asset values are large enough to provide security for the firm's debt.

Investors are interested in whether the assets are of the type and value to support the firm's debt and provide a residual for the equity holder. Comparative statements are more useful to creditors and investors because they reveal trends in the asset values and changes in the relationship between assets and liabilities. However, the statement of financial position has two major weaknesses. One, the assets are valued at historical cost in the statement, which may be significantly different from the current exit, market, reconstruction values. Two, the statement of financial position includes many values that are estimates, such as the depreciation of operational assets, the amortization of intangible assets, accruals, and the amounts of contingent liabilities. Generally, the statement of financial position, under the accounting concept of viewing the firm as a going concern, is considered to be less important than the income statement. This view of the statement of financial position supports the use of historical costs for valuing assets in this statement.[1]

THE INCOME STATEMENT (FORMERLY THE STATEMENT OF PROFIT AND LOSS)

Historically, the income statement has been the subject of controversy. Originally, it was argued that the income statement should reflect the results of the firm's operations. After all, it was argued, the firm is in business to manufacture automobiles, build buildings, and so on, and its success or failure at these objectives should be reported in the income statement. Income or loss from all other sources should be recorded in owner's equity. Because of the abuses in manipulating net income, this argument was rejected. Currently, the clean surplus (retained earnings) or all inclusive operating statement theory has been adopted by the accounting profession.

All costs and expenses with the exception of errors in prior periods financial statements must be reported in the income statement. The income statement can be separated into categories but the contents of the various categories are strictly defined. Income from continuing operations may be reported separately and then any losses from discontinued operations may be deducted. However, the discontinued operations must be a complete segment of a business; it cannot be discontinuance of a product or service. The result of this deduction is income from operations after discontinued operations.

[1]Glenn A. Welsch, Charles T. Zlatkovich, and Walter T. Harrison, Jr., *Intermediate Accounting* (Homewood, Illinois: Richard D. Irwin, 1979), pp. 137-143.

There may also be a section for unusual items, which must be unusual and not part of normal operations. After these items are deducted the result is income from operations after discontinued operations and unusual items. The next section is for extraordinary items and the cumulative effect of accounting changes. Accounting changes are a switch in inventory pricing such as a change in cost flow assumption from LIFO to FIFO. Extraordinary items must be unusual in nature and infrequent in occurrence. The deduction of these items results in income after extraordinary items and the cumulative effect of accounting changes. In all of these instances, the amounts should be shown both gross and net of income taxes. The applicable amount of income taxes must be shown with each category.

SECTIONS OF THE INCOME STATEMENT

The net income recorded in the income statement is the change in owner's equity from period to period caused by the firm's operations. The income statement typically has the following sections: (*a*) revenue which includes gross and net sales or gross and net project revenue for construction firms; (*b*) the cost of goods sold or the cost of sales for the construction firm (these are the matched costs, that is, the costs are matched with the revenue they produced); (*c*) the gross income or gross margin; (*d*) the expenses or period costs; (*e*) net income before taxes; (*f*) federal and state income taxes; and (*g*) net income from continuing operations after income taxes.

The construction firm's income statement differs from that of other firms in content rather than form. The period expenses may be less than they would be otherwise. Period expenses are deducted from revenue in the period in which they are incurred regardless of their relationship to the production of revenue. If a construction firm, however, can identify the expense to a particular project or contract or group of projects or contracts, it will charge the expense to that project or contract and include the expense in the project work in progress inventory and match the expense with the revenue produced.

The two basic forms of the income statement are the multiple step and the single step. The multiple step income statement is prepared in detail. In addition to including the total revenue and expenses, the statement lists the calculations required to convert gross revenue to net and the detail expense accounts as well as the calculation of cost of goods sold. The single step income statement is a summary report. Only the total revenue and expenses and the differences are included. In either case, the calculation of earnings per share is incorporated into the report. Exhibit 6 is an example of an income statement prepared in multiple step form.

THE IMPORTANCE OF THE INCOME STATEMENT

Early statements of profit and loss were prepared on a cash basis. Revenues were included in the statement when cash was received for the product or service and expenses were deducted or reported when paid. These statements were hard to interpret because of the failure to match revenue and expenses. Therefore, the profession developed the concept of net income and the accrual system. Net income is used as a surrogate for net cash inflow. Although it is an imperfect substitute, it is more likely that a firm that is profitable will generate a larger cash inflow than a similar firm that is not profitable over the long run (a long period of time).

Investors and creditors consider the income statement very important to their decisions. The earnings per share is considered one of the most important numbers on the income statement. When compared to past performance, earnings per share is the basis for the purchase or sale of the firm's capital stock. To the investor the income statement is the measure of management's performance for the accounting period.

Creditors are interested in whether or not the firm can support its debt load or potential debt load and still increase its wealth. The income statement is the statement which indicates whether the firm is performing well enough to meet the creditors expectations. In most cases, the success or failure of the going concern is reflected in the income statement.

THE STATEMENT OF RETAINED EARNINGS

The statement of retained earnings reconciles the beginning and ending balances in the retained earnings account in the owner's equity section of the statement of financial position. The retained earnings statement begins with the previous period's ending balance. Next, prior period adjustments are either added to or subtracted from the beginning balance. Then the net income for the accounting period is added and the distributions of capital (dividends on capital stock) are subtracted. The remainder is the ending balance in the retained earnings account. This balance is then further subdivided into appropriated retained earnings and unappropriated (available for the payment of dividends) retained earnings. A sample statement of retained earnings is included as Exhibit 7. The sample is illustrated as a separate statement; however, the statement of retained earnings may be included within and listed at the bottom of the income statement.

THE STATEMENT OF CHANGES IN FINANCIAL POSITION

The statement of changes in financial position is a relatively recent addition to the published financial statements. In the past this statement was elective; that is not true today! The statement of changes in financial position can be prepared on one of two bases. It may be prepared on the working capital or the cash basis. The most popular basis is the working capital basis. When the statement is prepared on the cash basis, the net result of the statement equals the change in the firm's cash balance for the accounting period. When the statement is prepared on the working capital basis, it can be prepared to balance to the change in working capital or the applications and resource sections can be balanced by the change in working capital from period to period. A statement of changes in financial position is included as Exhibit 8. The exhibit is prepared on both a working capital and a cash basis.

THE SECTIONS OR INCLUSIONS IN THE STATEMENT OF CHANGES IN FINANCIAL POSITION

Certain information must be disclosed in the statement of changes. The information can be disclosed in separate sections of the statement or it can be disclosed within the body of the source and application section. The three sections or inclusions are: (a) the sources and applications of funds separated into sources provided by operations, extraordinary items, and other (if the statement is prepared on a cash basis the source and application would be of cash rather than funds); (b) the changes in the total working capital and in the individual working capital accounts from accounting period to accounting period; and (c) changes in financing activities for the period, that is, if stock was exchanged for equipment, the exchange must be disclosed.

Usually, the financing activities are treated as if cash were received and then expended. In the case cited above, it would be treated as if cash were received from the sale of the stock and the cash were used to purchase equipment. And when the statement is prepared on a working capital basis, the largest single amount on the statement is the change in working capital; therefore, the statement is of little value without the schedule of changes in working capital accounts which explains the total change in working capital.

USES OF THE STATEMENT OF CHANGES IN FINANCIAL POSITION

The requirement that a statement of changes in financial position be prepared has already been noted. However, the requirement is specifically that a statement of changes must be prepared whenever an income statement is prepared. The income statement is prepared on an accrual basis with the matching of revenue and expense as its prime objective. Therefore, the income statement emphasizes net income or loss and does not tell the reader anything about cash flow or the use of total funds.

The statement of changes in financial position provides the user with this information. The reader can compare the flow of funds with the income flow recorded in the income statement. Thus the user has both sides of the picture, the flow and performance in the income statement and the performance for managing the funds (or cash) generated by operations and other sources. The two statements should be used together to completely evaluate the firm's performance for the accounting period.

The format of the statement of changes in financial position resembles that of the income statement for a more meaningful comparison. Funds provided by operations are listed first and then funds provided from other sources. The funds generated by extraordinary items are listed at the end just before the total funds provided. This order of the statement of changes supports comparability between the two statements.

For the construction firm, the statement of changes in financial position prepared on a working capital basis is of limited usefulness. The need for cash in the construction industry because of the lack of inventory to repossess, the tradition of working with others' funds, and the extensive use of accounts receivable retention make the cash basis more applicable. The statement of changes in financial position prepared on a cash basis with comparative data and comparable income statements would provide the necessary measure of the construction firm's ability to continue with its present policies and meet its obligations.[2]

[2]Donald E. Kieso and Jerry J. Weygandt, *Intermediate Accounting* (New York: John Wiley & Sons, 1977), pp. 112-134, 133-167, 903-983.

THE SELECTION OF A FISCAL PERIOD

The fiscal year is often taken for granted by the firm's managers and employees. However, the fiscal year is a matter of selection. The firm may select the fiscal period that best suits its objectives. As a general rule, the fiscal period is the firm's operating cycle or one year whichever is longer. In the nonconstruction industries, the one year fiscal period is the tradition and there are very few exceptions. Most construction firms also use the one year fiscal period; however, if a construction firm chose the completed contract method of reporting a strong argument could be made for the construction period as the operating cycle. Again, because of the need to compete for funds few construction firms use a reporting period of more than one year.

Once the fiscal period has been chosen, the beginning and ending dates must be selected. The calendar year should not be selected unless it supports the firm's operating objectives. One reason a fiscal year other than the calendar year might be selected is to support the planning cycle. If it were easier to predict based on a selling or construction cycle than the calendar year, a fiscal year should be chosen that supports the construction cycle!

However, the choice is more difficult for a construction firm. In many, if not most parts of the United States there are seasons when little, if any, constructing can be done. From the perspective of planning, selecting a fiscal year that begins before the construction cycle would allow time for the planning process and the preparation for operations. But this may not be the optimum alternative for the construction firm. It might be more profitable for the construction firm to have the fiscal period end during the nonconstruction or light construction months. For example, if the firm has a policy of retaining as many key employees as possible in the "off season" and the "off season" is at the beginning of the fiscal period, the plan must include sufficient volume and returns to support this personnel and provide a net income. The firm is, in effect, wagering on the future, that is, the next operating year.

The firm may, however, select a period with the "off season" at the end of the period. This is not the optimum choice to support planning, but if the operating year did not meet expectations the firm might not be able to retain or might not want to retain as many personnel as they otherwise would. They could adjust their costs to support their income objectives. The alternative they have available may not be the alternative they would prefer to select but it is at least factual rather than conjectural.

Another important influence on the selection of a fiscal period is the desire for comparability. A fiscal year may be selected to avoid comparability with competitors. Often the firms in one industry adopt the same fiscal year to facilitate comparability between firms. Because of the same reporting period they are able to trace each other's changes in assets, net income, and other financial ratios and compare them to their own. If the industry and the fiscal period are the same, the firms can compare quarterly as well as annual data. From this information a firm may be able to detect a competitor's long and short run tactics and strategies. Realizing that a competitor can apply the same techniques to the firm's financial statements, the firm may choose a fiscal year that is different from the rest of the industry. This choice will make it more difficult for competitors to deduce the firm's plans.

The selection process does not end with the selection of the fiscal year. If the calendar year is selected as the fiscal year, will the calendar month be used as the basis for interim reporting? If a fiscal year rather than a calendar year is selected, should the calendar month be used for interim reporting within the fiscal year? Other alternatives are to (a) select a 30 day month for interim reporting with any remainder of the year included in the last month; (b) divide the fiscal year into 13 months with the excess day or two falling consistently in the thirteenth month; and (c) use a four week, four week, five week quarter for each quarter, with any remainder included in the last quarter.

The criterion for selecting the interim reporting interval is the importance to the firm of comparability between months and quarters. If comparability between interim reports is vital to performance, an alternative other than the calendar month should be selected. If consistency and comparability between interim periods is relatively unimportant, the calendar month will suffice.[3]

SUMMARY

In this chapter the statement of financial position, the income statement, the statement of retained earnings, and the statement of changes in financial position are reviewed. These are the four statements that are included in the firm's annual report. Interim reports may include all of these statements or only the statement of financial position and the income statement. A discussion of these statements is not complete without a discussion of the selection of

[3]William W. Pyle, John Arch White, and Kermit D. Larson, *Fundamental Accounting Principles* (Homewood, Illinois: Richard D. Irwin, 1978), p. 121.

the fiscal period (reporting period) because of its effect on the content
and comparability of the financial statements. Therefore, the criteria for selecting a fiscal period are examined.

**EXHIBIT 6. Income Statement (in Multiple Step Form)
for the Year January 1, 1980, through December 31, 1980**

Description	Amount	Amount
	(In thousands)	
Revenue from construction		
Gross project revenue	$100,000	
Less: credits and adjustments	10	
Net revenue from construction projects		$99,990
Cost of revenue		
Beginning inventory of raw material	$ 1,500	
Raw material purchases	15,000	
Less: ending raw material inventory	−1,000	$15,500
Beginning value of construction in progress	$ 11,000	
Construction costs		
Direct project labor	9,000	
Indirect project labor	3,000	
Supplies used in construction	100	
Equipment charges to projects	200	
Truck hire	50	
Project supervision	250	
General and administrative costs		
Identified to projects	$ 60	
Subcontracting costs	10,000	
Total costs of construction in progress without raw material	$ 33,660	
Less: construction in progress at the end of the period	560	$ 33,100
Total cost of revenue		$ 48,600
Gross revenue (margin)		$ 51,390
Expenses (Not Identified to projects):		
Selling: contract administration-central	$ 1,000	
General and administrative		
Accounting	$ 200	
Personnel	75	
Depreciation	60	
Taxes (other than income)	1,000	
General	1,265	$ 3,600

EXHIBIT 6. (Continued)

	Amount	Amount
	(In thousands)	
Net income from continuing operations before income taxes and other items		$ 47,790
Other income and expense		
Interest expense	$ 1,000	
Interest income from accounts and notes receivable plus income from back charges	100	$ 900
Net income from continuing operations before income taxes		$ 46,890
Less: federal, state, and subdivision income taxes applicable to continuing operations		22,890
Net income from continuing operations after income taxes		$ 24,000
Unusual loss from sale of New York plant	$ 2,000	
Less: applicable income tax	1,000	1,000
Net income from operations after income taxes and unusual items		$ 23,000
Discontinued operations		
Loss from sale of concrete division	$ 10,000	
Less: applicable income tax	5,000	5,000
Net income from operations after income taxes, unusual items, and discontinued operations		$ 18,000
Extraordinary item—loss from flood	$ 6,000	
Less: applicable income tax	$ 3,000	$ 3,000
Cumulative effect on prior periods of retroactive application of new depreciation method	$ 4,000	
Less: applicable income tax	2,000	$ 2,000
Net income after income taxes, the cumulative effect of accounting changes, and extraordinary items		$ 13,000
Equivalent number of common shares outstanding		10,000,000

Earnings per Share

Net income from continuing operations after income taxes	$2.40
Unusual loss	(.10)

EXHIBIT 6. (Continued)

	Amount	Amount
	(In thousands)	
Net income from operations after income taxes and unusual items		$2.30
Discontinued operations		(.50)
Net income from operations after income taxes, unusual items, and discontinued operations		$1.80
Extraordinary item		(.30)
Cumulative effect of accounting changes		(.20)
Net income after income taxes, the cumulative effect of accounting changes, and extraordinary items		$1.30

EXHIBIT 7. Statement of Retained Earnings for the Year Ended December 31, 1980

	Amount	Amount
Description	(In thousands)	
Retained earnings January 1, 1980		$200,000
Less: prior period adjustment—accumulated depreciation on equipment was in error in 1979 financial statements—accumulated depreciation was understated		(1,000)
Adjusted retained earnings at January 1, 1980		$199,000
Add net income for the 1980 fiscal year		13,000
Retained earnings before distributions to owners		$212,000
Deduct		
Dividends declared on preferred shares	$ (5,000)	
Dividends declared on common shares	(27,000)	(32,000)
Retained earnings at December 31, 1980		$190,000
Retained earnings appropriated at December 31, 1980		(90,000)
Unappropriated retained earnings at December 31, 1980		$100,000

Note: A similar statement can be prepared for single owner firms and partnerships.

74

EXHIBIT 8. Statement of Changes in Financial Position for the Year Ended December 31, 1980

Working Capital Basis (1)

Description	Amount	Amount
		(In thousands)
Resources provided by		
Operations		
Net income before extraordinary items		$165,000
Add expenses not requiring an outlay of working capital in the current period		
Loss on sale of equipment	$ 25,000	
Depreciation expense	20,000	
Amortization of patent	10,000	55,000
Resources provided by operations before Deduction of nonoperating gain		$220,000
Deduct nonoperating gain included in net income which does not provide working capital in the current period		(30,000)
Working capital provided by operations		$190,000
Extraordinary gain (net of income taxes)		100,000
Working capital provided by operations and extraordinary gain		$290,000
Other sources of funds		
Sale of investment in common stock		$ 50,000
Proceeds from sale of equipment		25,000
Sale of building for cash—book Value—extraordinary gain on sale is included above		150,000
Total resources provided by operations and other sources		$515,000
Resources applied to		
The purchase of equipment		$ 50,000
An increase in working capital accounts		465,000
Total resources applied		$515,000

Changes in working capital accounts

Description	1979	1980	Increase	Decrease
		(In thousands)		
Cash	$ 20,000	$ 50,000	$ 30,000	
Marketable securities	40,000	150,000	110,000	

EXHIBIT 8. (Continued)

Description			Amount	Amount
			(In thousands)	
Notes receivable	25,000	125,000	100,000	
Accounts receivable	$120,000	$340,000	$220,000	
Inventories	300,000	405,000	105,000	
Notes payable	75,000	100,000		$ 25,000
Accounts payable	100,000	175,000		75,000
Changes in working capital accounts			$565,000	$100,000
Net increase in working capital included above				465,000
Proof of balance of working capital accounts.			$565,000	$565,000

Financing Activities

Description	Amount (In thousands)
Issued a long term note payable for equipment	$ 12,000
Sold and issued common stock to retire a long term mortage payable	4,000
Total financing activities	$ 16,000

Statement of Changes in Financial Position—Cash Basis

Description	Amount	Amount
	(In thousands)	
Sources of cash		
Operations		
Net income		$ 34,000
Add		
Depreciation expense (noncash item)	$ 1,082	
Loss on sale of equipment		
Not requiring cash	37,604	38,686
Less: payments on trade payables		(17,612)
Cash provided by operations		$ 55,074
Other sources of cash:		
Issue of long term debt		$100,000
Issue of common stock		15,000
Sale of equipment		26,000
Sale of short term securities		6,926
Total cash provided		$203,000

EXHIBIT 8. (Continued)

Description	Amount	Amount
	\(In thousands\)	
Uses of cash		
Retirement of long term note		$ 34,000
Payment of mortgage		$ 5,000
Payment of preferred and common dividends		27,000
Purchase of equipment		127,000
Total cash used		$193,000
Net increase in cash		$ 10,000

Changes In Working Capital Accounts

Description	1979	1980	Increase	Decrease
		(In thousands)		
Cash	$ 20,000	$ 40,000	$ 10,000	
Marketable securities	40,000	170,000	130,000	
Notes receivable	25,000	125,000	100,000	
Accounts receivable	120,000	340,000	220,000	
Inventories	300,000	405,000	105,000	
Notes payable	75,000	100,000		$ 25,000
Accounts payable	100,000	175,000		75,000
Changes in working capital accounts			$565,000	$100,000
Net increase in working capital				465,000
Proof of balance of working capital accounts			$565,000	$565,000

Financing Activities

Description	Amount (In thousands)
Issued common stock to retire long term bonds	$ 60,000
Assumed mortgage to purchase plant	200,000
Total financing activities	$260,000

NOTES:
(1) Donald E. Kieso and Jerry J. Weygandt, *Intermediate Accounting* (New York: John Wiley & Sons, 1977), pp. 965-970, 1009-1010.
(2) "Professional Notes," *The Journal of Accountancy*, December 1979, pp. 88-96.

QUESTIONS

1. Why does the balance sheet (statement of financial position) have an "as of" date?
2. Describe the classification of the asset side of the balance sheet.
3. Are there any peculiarities in a balance sheet for a construction firm?
4. How many formats are there for preparing the statement of financial position? Describe each format.
5. What are some of the uses of the statement of financial position?
6. What are some of the weaknesses of the statement of financial position?
7. How was the controversy concerning what was to be included in the income statement resolved?
8. How many formats are there for preparing the income statement? Describe each format.
9. Is the income statement a surrogate for cash flow? Explain.
10. In detail, list and discuss the various sections of the income statement and the reporting of earnings per share.
11. To whom and why is the income statement important?
12. Define the statement of retained earnings.
13. Describe the content of the statement of retained earnings.
14. What is the purpose of the statement of changes in financial position?
15. What are the two bases for preparing the statement of changes in financial position?
16. Describe the three types of information that must be included in the statement of changes in financial position.
17. When is a statement of changes in financial position prepared and what are its uses?
18. List and describe the criteria for selecting a fiscal year. What is the general rule?
19. Describe some of the commonly used fiscal months and discuss the criteria for choosing between them.

REFERENCES

Anthony, Robert N., and James S. Reece, *Management Accounting* (Homewood, Illinois: Richard D. Irwin, 1975).

Chorba, George J., *Accounting For Managers* (New York: American Management Association Extension Institute, 1978).

Gordon, Myron J., and Gordon Shillinglaw, *Accounting: A Management Approach* (Homewood, Illinois: Richard D. Irwin, 1974).

Horngren, Charles T., *Cost Accounting: A Managerial Emphasis* (Englewood Cliffs, New Jersey: Prentice-Hall, 1977).

Kieso, Donald E., and Jerry J. Weygandt, *Intermediate Accounting* (New York: John Wiley & Sons, 1977).

Meigs, Walter B., A. N. Mosich, and E. John Larsen, *Modern Advanced Accounting* (New York: McGraw-Hill Book Company, 1979).

Montgomery, A. Thompson, *Managerial Accounting Information* (Menlo Park, California: Addison-Wesley Publishing Company, 1979).

"Professional Notes," *The Journal of Accountancy*, December 1979.

Pyle, William W., John Arch White, and Kermit D. Larson, *Fundamental Accounting Principles* (Homewood, Illinois: Richard D. Irwin, 1978).

Rayburn, L. Gayle, *Principles Of Cost Accounting With Managerial Applications* (Homewood, Illinois: Richard D. Irwin, Inc., 1979).

Welsch, Glenn A., Charles T. Zlatkovich, and Walter T. Harrison, Jr., *Intermediate Accounting* (Homewood, Illinois: Richard D. Irwin, 1979).

CHAPTER 6

INVOICING, ACCOUNTS RECEIVABLE, AND CASH RECEIPTS

Controlling invoicing, accounts receivable, and cash is difficult in a manufacturing environment; however, in the construction industry, with its diverse locations and rush to complete each project, control is extremely hard to achieve. Without the support and cooperation of field personnel, it is almost impossible! But, because of the continuous need for cash, which is characteristic of the construction industry, successful control of these functions is a vital necessity.

INVOICING CUSTOMERS

If retailing is excepted, it is an axiom that customers will not pay for goods and services unless they are billed, that is, receive an invoice for the goods or services received. In construction, a product is not shipped with the necessary paperwork. Shipping specialists are not available to protect the firm's interests. The construction industry, in general, relies on the field personnel to notify the financial group that the project is complete. The problem often is not that notification is not received, but it is not received on a timely basis. The problem is timing. The faster the billing department is notified, the faster the invoice can be prepared and the cash collected. Again, there is heavy reliance on field personnel.

Most construction projects are performed under a contract. The contract contains provisions for the invoicing of the customer (owner) and, therefore, affects the construction firm's cash inflow. The contractual provisions determine when invoices may be submitted; the extent or the percentage of the work completed that may be invoiced; the amount of progress payments permitted, if any; and the amount or percent of retention that will be withheld and from what type of invoices the retention will be withheld.

It is not uncommon for a contract to allow a billing when each phase of the project is completed, as defined in the contract. However, the contract may require that only a specified percent of each phase be billed until the total project is completed. This limitation is in addition to the retention withheld. The contract may also specify that the construction firm can receive progress payments as each item within a phase of the project is completed up to the total allowed to be invoiced for that phase, which may be subject to the limitation noted above.

In the construction industry the withholding of a retention until the final product is accepted by the customer is traditional. The retention is usually withheld from each invoice submitted by the construction firm except the final invoice. The percent of retention withheld varies from 5 to 15%, with 10% being the most common. There is, however, a trend to eliminate, reduce, or permit early billing of the retention for superior performance on the part of the construction firm. Usually superior performance is interpreted to mean performance that is within the contract cost objectives and specifications but exceeds (betters) the contract date of completion requirements.[1]

One contract provision that can and usually does cause a delay in receiving payment is the detailed invoicing provision. This part of the contract includes the detailed information to be printed on the invoice. In some cases, the contract specifies that invoicing will be done on forms furnished by the customer. If these provisions are not precisely followed the invoice will be returned to the construction firm unpaid. Thus a delay in cash inflow. Because of the perennial shortage of cash in the construction industry, exact adherence to these contract requirements can prevent customers from withholding and using the construction firm's cash.

THE NEED FOR INPUT FROM FIELD PERSONNEL

In the last few paragraphs the various invoicing possibilities were discussed. But the financial division of the firm must first be aware of the completion of the various stages of the work on each project. Unlike manufacturing, there are no delivery specialists. The firm must rely on the field personnel. One reporting technique is the

[1]William W. Pyle, John Arch White, and Kermit D. Larson, *Fundamental Accounting Principles* (Homewood, Illinois: Richard D. Irwin, 1978), pp. 177-220.

submission of measurement reports by field personnel either continuously or at specific periods throughout the month. Measurement reports usually specify the amount of physical work completed, based on an engineering measurement, up to the time of measurement. The statistic is usually stated as 100% completion or the percent of completion of that project phase. The financial department subtracts the percent invoiced at the last measurement from the current percent and submits an invoice to the customer. The finance department also checks the contract billing provisions for form, project phase billing limitations, and the amount that must be shown as retention.

Sometimes the contract specifies that invoices are to be submitted upon the completion of certain milestones such as the completion of the foundation of a building or the completion of the interior support structure. When these milestones have been completed the financial department should receive a progress report from field personnel and the financial department can submit an invoice to the customer. The invoice must be checked against the contract billing provisions before submission to the customer.

Completion reports specify the percentage of completion of a construction project at the report date. These reports are used for complex projects where a simple measurement of the work completed is not possible. Field personnel, with the agreement of customer personnel, estimate the extent of completion of the project at the report date. Both the construction firm's personnel and the customer's representative sign the report and it is submitted to the construction firm's financial department for preparation of the invoice.

If the construction firm sells a product such as asphalt, concrete, or gravel, evidence of the delivery of the product is necessary. In these cases a delivery ticket is prepared and signed by the customer or his representative. In most cases the plants from which the product is shipped are highly automated and the delivery ticket is prepared automatically at the time of shipment. Regardless of how the receipt is prepared, a copy must be sent to the financial area for the proper and timely billing to the customer.

CONTROLLING INVOICE PREPARATION

The reader should recall that in Chapter 1 the journal entry for recording accounts receivable included a credit to sales on contract or project revenue. Therefore, the recording of the receivable affects the reported construction revenue. Almost all revenue in construction is from invoices, that is, cash sales are either nonexistent or extremely small. To record accurate sales revenue for the period, the invoices to customers must be accurately recorded. It is not only important for cash flow to record and prepare invoices on a timely basis, it is also important to record all but not duplicate revenue from construction. Because net income is a function of revenue less costs, all invoices must be recorded but revenue (sales) must not be overstated.

One method for controlling invoice preparation is the requirement that invoices be prepared only from a document that is evidence of completion of the work required or an adjustment to a previous invoice. Documents that are evidence of the completion of work should originate with field personnel. Adjustment invoices should be supported by an internally originated document that has the proper approval, indicates the reason for the adjustment, and includes the calculation of the amount of the invoice.

Another method for controlling the preparation of invoices is the use of a sequential numbering system. The invoice number can include the fiscal year, the contract number, and the customer number, but a portion of the number should be sequential for the fiscal period. The sequential number can be used to verify that all assigned invoices were issued and to ascertain that all invoices that were issued were recorded.

Although it is possible with an invoice numbering system to not prepare an invoice, this possibility can be reduced with a system of double checking. One employee can be assigned the responsibility for matching invoices and contracts or delivery tickets. This employee could then inquire about delivery tickets without invoice numbers and contracts that have not had any invoicing activity for a long period of time. In addition, a manager should be assigned to occasionally trace invoices to contracts and vice versa. Both the internal and external auditors will test the invoicing system. With these precautions it is unlikely that material errors will occur in the invoicing function.

INVOICE INFORMATION

In addition to the invoice number, date, contract number, customer name and address, delivery location, and approval, the invoice must indicate the total billed to date, less the amount billed on the previous invoice, the result being the amount billed this period. The invoice must also show the total amount of retention withheld by the customer through the invoice date. The invoice must specify the work done or product delivered, the construction project number, and the phase of the project for which the work was performed and if necessary must be prepared in the form required by the contract and perhaps on the forms specified. It is important that the invoice information be precise and conform to the contract provisions. Usually the more exact and meaningful the invoice the more rapidly payment will be received. It is certain, however, that invoices that are poorly prepared will delay payment and adversely affect the firm's cash inflow.

Progress billings are billings that are allowed while the work is in process and before the project or a phase of it is complete. An invoice for a progress billing must be labeled as such. The contract specifies the terms for a progress billing. The billing is usually a percent explicitly stated in the contract of a phase or a milestone of a construction project. The purpose of a progress billing is to provide the construction firm with interim financing. The construction firm is usually in the position of working on projects that require a long period of time before completion. The suppliers to the firm require payment, in most cases, long before the construction project is completed. Progress payments reduce the cash squeeze on the construction firm. Even though progress billings are not specified in the contract, it may be possible to submit them. Progress billings are a tradition in the construction industry and are the rule rather than the exception. But, for the protection of the construction firm, progress billings should be included in the contract as one of the performance provisions.

Final billings to customers are made after the acceptance of the work. The final billing to the customer includes the work done since the last invoice and usually a rebilling of the retention withheld by the customer over the life of the project. Final billings are labeled as such. At this point all disputes between the customer and the construction firm must be resolved. If the construction firm has been resolving differences with the customer as the firm becomes aware of them, the final invoice will be processed rather rapidly. If, however, the disputes have waited until submission of the final invoice, that invoice will not be paid until all disputes are resolved. The delay in resolving the disputes with the customer will delay payment of the final invoice and thus adversely affect the construction firm's cash inflow.

It is unlikely that the customer (owner or general contractor) will pay the retention withheld without a rebilling from the construction firm. There is no doubt that the customer will check the rebilling against his records. Therefore, the construction firm must maintain accurate records of the retention withheld. To expedite the processing of the invoice the retention should be listed by invoice number and date of the original billing. This will make the customer's review easier and avoid the need for a request for additional information. Construction firm personnel must maintain adequate records to avoid unnecessary delays in the firm's cash inflow.

ACCOUNTING TREATMENT

The accounting journal entries for the recording of billings are similar for construction and nonconstruction firms. The major differences are in the credit side of the entry, that is, the account title of the credit, the type of customers, and the uniqueness of the product. In general, the credit is classified by type of customer and the contract number, but both the debit and credit can be as detailed as needed to accommodate the information requirements of the construction firm. The general type entries are:

Debit: Accounts receivable—contract #XXX—progress billing $XX,XXX.XX. Or
Accounts receivable—contract #XXX—final billing $XX,XXX.XX. Or
Accounts receivable—contract #XXX—retention $XX,XXX,XX.
      Credit: Project revenue—contract #XXX—progress billing $XX,XXX.XX. Or
Project revenue—contract #XXX—final billing $XX,XXX.XX. Or
Project revenue—contract #XXX—retention $XX,XXX.XX.

For a specific firm, the billing may contain additional information:

Debit: Accounts receivable—contract #XXX—progress billing—federal government $XX,XXX.XX
      Credit: Project revenue—contract #XXX—progress billing—federal government $XX,XXX.XX

Or:

Debit: Accounts receivable—contract #XXX—final billing—state government $XX,XXX.XX
      Credit: Project revenue—contract #XXX—final billing—state government $XX,XXX.XX

Or:

Debit: Accounts receivable—contract #XXX—retention—private contract $XX,XXX.XX
    Credit: Project revenue—contract #XXX—retention—private contract $XX,XXX.XX

The accounting entries can be detailed further into types or categories of private contracts and the government classification can be further classified by subdivisions such as county and town or city governments.

Regardless of the number and detailed classifications used for the recording of the invoice, the sales and revenue entries must be detailed by construction project. Recording the project revenue by project is necessary for the preparation and analysis of net income by project, which are discussed in more detail later in this book. Without this information the analysis of project profits is not possible and the firm will not know which of its managers or efforts were successful. Therefore, the accounting entry should include, in addition to the information above, the following:

Credit: Project revenue—contract #XXX—retention—private contract $100,200.00 (amount assumed) Project detail:

Project #1234	$ 25,100.00
Project #1235	25,100.00
Project #1236	20,000.00
Project #1237	30,000.00
Total	$100,200.00

ACCOUNTS RECEIVABLE

The proper administration of accounts receivable is important in any firm, but in the construction industry it is particularly important. The customer must not be given an excuse to withhold payment of an invoice. Improper accounts receivable are often the excuse! Control of accounts receivable requires the use of control accounts and a detailed subsidiary ledger for each accounts receivable control account. The general ledger would include an accounts receivable control account for each contract such as: accounts receivable—final billings; accounts receivable—progress billings; and accounts receivable—retentions—contract #XXXX. For each of the control accounts in the general ledger there must be a subsidiary ledger that equals the control account. The double posting of the amounts to both the control account and the subsidiary ledger is a method of verifying the postings because the balances of the control account and its subsidiary ledger must equal at all times.

How many subsidiary ledgers there are for each control account is a function of the degree of summarization in the control account. The usual approach is a control account that includes all progress billings. This account would have two subsidiary ledgers. One subsidiary ledger would be organized by customer and contract within customer and the other subsidiary ledger would be by project. In each case the subsidiary ledger would equal the balance in the accounts receivable control account. An illustration of an accounts receivable control account and its subsidiary ledger is included as Exhibit 9.

CUSTOMER DEDUCTIONS

In the construction industry cash discounts are rare. Cash discounts are used when a construction firm sells a product such as aggregates or concrete. However, returns are impossible! These two facts would appear to make the maintenance of receivable records less complex in the construction industry. But that is not the case. General contractors and owners (customers) withhold many amounts from invoice payments and in the process usually delay the payment of invoices. Amounts are withheld for retention, for not conforming to contract specifications (cracks, not proper color, leaks), and for late completion or for delaying the next phase of completion or the next subcontractor.

Regardless of the reason given for the withhold, the cause must be resolved quickly. Unresolved customer deductions can cause delay in payment of invoices by customers, and the longer a deduction is unresolved the less chance the construction firm has of ever collecting the amount withheld. Apparently, the assumption is that the failure of the construction firm to resolve a deduction is implicit approval.

Retention is traditional in the construction industry. From every invoice submitted the customer, regardless of the type of invoice, will withhold a percentage as retention. The retention will not be paid until the job is accepted by the customer and all disputed items have been resolved. Although the exact requirements for the retention are included in the contract, the percentages usually withheld are 5 or 10%. Even though it is unusual, there are contracts in which the customer has waived his (or her) right to a retention.

One of two procedures are generally used to invoice a customer for retention. One is to submit all invoices net and to bill for the retention at the final billing. This method recognizes the delay in collecting the amounts withheld and does not result in an outstanding account receivable for the retention during the construction period. It also avoids the correction of revenue for customer deductions which are resolved in the customer's favor.

The second method is to invoice the customer for the gross amount and to carry the retention withheld as an account receivable until it is paid. The advantage of this method is that revenue may be recognized when the invoice is submitted although payment may not be received for the retention until contract completion. The major disadvantage of this method is that revenue may be overstated by the amount of customer miscellaneous deductions that are allowed. The second method is the most widely used one. With both methods, the total retention is usually resubmitted to the customer on one invoice at contract completion. Again, the payment of the invoice by the customer can be delayed through disagreement about the amount of retention withheld and by the failure to resolve customer miscellaneous deductions.

One function within the construction firm that affects the payment of invoices by the customer is cash application. Cash application is the application of the customer's payment to the proper outstanding invoice. When customers include remittance advices, the payments and deductions can be identified to the proper invoices. When the customer pays on account or submits payments without any identification, cash application is most difficult. Unless amounts minus deductions can be traced to specific invoices, the payments are applied to the oldest (longest outstanding) invoices first. Because most firms send their customers monthly statements of amounts due, the risk in this approach to cash application is that the customer and the construction firm may not apply the payments similarly. Thus disputes arise concerning the amounts due on specific invoices that delay future payments. The solution is to convince the customer to furnish remittance advices or to respond quickly to resolve disputes. Because of the need for cash, it may not be in the customer's interest to resolve these disputes quickly.

THE AGING OF ACCOUNTS RECEIVABLE

The administration of accounts receivable would be incomplete without the reporting and the analysis of the age of the receivables. It is a truism that the older the balance the less the chance of collection. In addition to its use as a monitor of collections, the aging is used to compute late payment penalties. However, in the construction industry late payment penalties are rarely used or included in construction contracts unless a product is involved.

For construction firms that do not sell a product, the aging is useful for following up on unpaid invoices. This can be a clue to unresolved deductions or other problems. The aging, as described and illustrated here, can be done on a computer or calculated manually. Regardless of the available techniques, the objective is to categorize the receivables balances by customer and days past due. The aging can be subtotaled by type of receivable such as progress billings, final billings, and retention. Past agings can be compared with the current aging to identify trends. The aging report can be distributed throughout management for review and follow-up.

The format for the accounts receivable aging should include balances that are current as well as those that are past due. The final total of the aging should equal the total of the accounts receivable. The aging categories can be designed for the particular firm but should at least include the following: (*a*) current amounts; (*b*) amounts 1 to 30 days past due; (*c*) amounts 31 to 60 days past due; (*d*) amounts 61 to 90 days past due; (*e*) amounts 91 to 120 days past due; (*f*) amounts 121 to 150 days past due; (*g*) amounts 151 to 180 days past due; and (*h*) amounts that are 181 plus days past due. An example of an accounts receivable aging is given in Exhibit 10.[2]

CASH RECEIPTS

The most effective method for controlling the receipt of cash is not to receive any. Regardless of the effort made by firms to receive payments by check, however, cash will enter the firm through its employees and through the mail. Cash sales are rare in the construction industry. Firms that sell products such as concrete and aggregates receive more cash than those that sell construction services, but even they receive cash. The cash comes from employees who purchase small items from the firm, such as gasoline, tools, and pieces of material; it also comes from employees who receive cash from customers and customer employees for doing small cleanup and patching jobs; for sales to customer employees; and for purchase of special licenses and payment of special fees such as hauling fees; and cash arrives through the mail for payment of the items listed above plus payments for small invoices submitted to customers.

The problem for the construction firm is to decide whether to establish control procedures for the cash received. The firm's managers must judge whether the controls will consume more cash than will be lost through the lack of controls. If the managers decide

[2]*Ibid.*, pp. 221-262.

that controls are necessary, proved techniques should be introduced. One of the weak points in cash control is when the cash enters the firm. Although none of the methods employed is foolproof, an attempt should be made to establish and enforce control at this point. One control method is to require that employees give the customer a receipt and surrender a copy of the receipt with the cash. Another procedure is to require that the employee receiving the mail list cash received by customer and amount. In either case, the employee depositing the cash on the construction firm's behalf must not be the same employee receiving the cash. This will support a matching of the cash received versus the cash deposited.

An additional method of verification is checking the sales amount against the cash received. Sales can be verified by adding the delivery tickets for that customer and matching the total with the cash received. If the amounts are not equal a verification of the balance can be sent to the customer. An excellent control device is the continuous verification of receivable balances with the customer. Verification can be through monthly statements, employee contacts with customers, and contacts with the customer's accounts payable section in an attempt to reconcile the receivable and payable balances.

CONTROL OF CHECKS RECEIVED

Despite the objective of receiving payment in checks rather than cash, controls cannot be dispensed with. Checks must be listed upon receipt. After they are listed, they should be copied and deposited by an employee other than the employee who received them. The copies are used for the application of the payments to the customer's account. If a lock box system is used, a listing of the checks and copies are available from the correspondent bank. In addition, the funds become available to the construction firm more quickly and the information needed for the application of cash is available. Again, the purpose is to prevent employees from misapplying payments and compromising the funds and to prevent the checks from being stolen and the proceeds being misappropriated.[3]

CREDIT POLICIES

Cash flow in the construction industry is not an unsolvable problem. Construction firms must turn their cash quickly, even in the smaller

[3]*Ibid.*, pp. 263-296.

sections of the industry, where resources are committed for a relatively long period of time. Despite progress payments, retentions are withheld by customers. There are always disputes and other customer deductions. To properly manage any firm, including a construction firm, the managers must predict cash inflow within some reasonable range. Without this ability it is difficult to arrange bank financing when needed—managers will not know when it is needed! Aggressive administration of the credit function is one method for improving and making cash flow more uniform. Generally, credit policies can be classified as either preventive, corrective, or punitive.

PREVENTIVE POLICIES

These policies are enforced before the sale or before the construction contract is signed. The objective of these policies is to identify the poor paying customers before the contract is signed. One method for executing this policy is to check the construction firm's own files for a history of the customer's past payment record. If none is found, the construction firm can check with another construction firm with which it is familiar to determine whether they have any information on the customer. Or the construction firm's credit manager can inquire of the local credit manager's organization to identify a firm that knows the customer.

If no information is available from these sources or the construction firm would like more information, a credit agency that sells this type of information can be contacted. In addition, the firm's credit personnel can check the credit reference manuals and the financial press for customer information. The construction firm can also request that the customer furnish financial information before any agreement is signed.

Among the preventive policies are the contractual payment provisions and the terms of sale (payment). The contract payment provisions must specify when the construction firm may invoice the customer and when the customer will honor the invoices. There should also be a statement about what recourse the construction firm has if the customer violates the contractual payment provisions. The terms of payment are part of the credit policies because reasonable terms may result in receipt of payment within those terms. If, however, the terms of payment are unreasonable, payment may not be expected within the terms. If the terms are other than those that are traditional within the industry, enforcement of the terms may not be possible. In the construction industry early payment discounts are unusual and governments usually do not pay invoices for 60 to 90 days from receipt.

An additional protective device is personal guarantees. The construction firm may require the officers and/or owners of the customer to pledge their personal assets for the obligations of the customer. Through this device the officers and/or owners of the customer guarantee the payment of the customer's obligation to the construction firm. If the customer does not pay, the construction firm has recourse to the personal assets pledged by the officers and/or owners of the customer. This can be an effective device for protecting the construction firm from a cash loss.

CORRECTIVE POLICIES

One effective corrective policy is the imposition of a late payment penalty if the payment is not received from the customer within the agreed upon time period. Usually the penalty is assessed as a percentage of the past due balance for the period the balance is past due. A difficulty with this approach is that some states have laws that specify the maximum amount an individual may be charged for interest or penalty. This prohibition usually does not apply to a corporation; however, it may take a considerable expense to verify whether the customer is a proprietorship (single owner), a partnership, or a corporation. In addition, the various state laws may require that the customer agree to the penalty in writing. If the construction work performed includes services, the contract must provide for the imposition of the penalty. And if the customer is a government, it will be impossible to impose the penalty. Therefore, the contract price should include the cost of carrying a receivable from the government for at least 90 days.

Another corrective policy is the use of telephone calls, letters, and other reminders to the customer that the payment is past due. A typical approach is to first contact the customer by telephone, and if there is no response to follow the call with a letter reminding the customer that the payment is overdue. If the customer still does not respond more letters are sent and the tone of each successive letter becomes more severe. Finally, the collection effort will shift from corrective to punitive.

If the construction firm has material amounts of government billings the best procedure is to maintain continuous telephone contacts with the government disbursing office to resolve any procedural or technical difficulties with the invoice. If the firm is highly dependent on government contracts, the best procedure is to maintain continuous personal contact with the government disbursing office. It is likely that payment of an invoice by a government will be delayed because of a simple clerical error in the invoice. Usually the government disbursing office will not notify the contractor at the time of discovery of the error, but, instead, will reverse the disbursing process and return the invoice to the contractor. Through continuous contacts with the disbursing office the construction firm will become aware of the problem earlier, can prepare a corrected invoice, and reduce the delay in payment.

PUNITIVE POLICIES

When all else has failed the construction firm must turn to punitive measures to collect the amounts due it. Available procedures peculiar to the construction industry include the execution of the construction firm's rights under payment bonds or filing a claim under the mechanic's lien laws of the various states. The details of bonding and the lien laws are discussed later in this book. However, in general, the effect of filing a claim under a payment bond is to be reimbursed by the bonding company (usually an insurance company) and forcing the bonding company to file a claim for reimbursement against the insured (the construction firm's customer). In some cases, merely a threat to contact the bonding company or a call to the bonding company relating that the construction firm has not been paid by the customer will result in immediate payment.

The filing of a claim under a mechanic's lien law will give the construction firm a right to some of the proceeds from the sale of the real property on which the construction was performed. The property will not be free of encumbrances until the lien is extinguished. As with the payment bond, often the threat to file a lien or, as is permitted in some states, the filing of an intent to file a lien, will result in immediate payment.

Other procedures available to the construction firm are the use of a collection agency and/or the filing of a lawsuit. Because collection agencies have developed techniques that are useful and peculiar to their business, they can often get results when the construction firm's efforts have failed. However, as with all punitive efforts there is a cost to the construction firm. In this case, the collection agency usually retains a percent of the amounts collected. The collection agency will also initiate a suit on behalf of the construction firm. Some construction firms have their own legal staffs and prefer to initiate their own litigation. They will instruct the collection agency to notify them if their collection effort fails. The construction firm will then take the necessary legal action. It should be noted that corrective and punitive policies are both expensive and time consuming. The result is a reduced and delayed cash inflow to the construction firm and usually loss of a customer.

BACK CHARGES

Back charges are peculiar to the construction industry. Although in manufacturing and merchandising industries there are returns and allowances, they are less difficult to control than back charges in the construction industry. Again, the widespread geographic operations of the construction firm and the pressure to complete the construction project on schedule contribute to the problem. If the project is located in an area with changing seasons, the pressure for completion may be severe to avoid the delays often caused by onset of winter.

WORK DONE FOR SUBCONTRACTORS

There are many types of services a construction firm may perform for its subcontractors. Usually these services are included in the negotiation of the subcontract. But the unexpected can occur. The subcontractor may encounter a problem for which there is no equipment on site. In addition, the subcontractor may need the services of an expert for a short period of time or may be behind work schedule and leave the project site without performing the necessary cleanup. Some of the issues that arise because the construction firm performs these services for its subcontractors are as follows. (a) Does the subcontractor's representative have the authority to commit the subcontractor to reimbursing the construction firm for the performance of these services? (b) What evidence is available to support the proposition that the subcontractor or his representative requested the services? (c) What are the costs to the construction firm of collecting for these services from the subcontractor versus absorbing the costs? (d) What are the costs of risking delay in completion of the construction project versus forcing the subcontractor to perform these services himself or to formally commit himself to reimbursing the construction firm for these services?

In most cases, the communication concerning the need for these services by the subcontractor is between the construction firm's field personnel and the subcontractor's representative at the project site. From the perspective of financial control, the construction firm's field personnel should have the subcontractor put his request in writing and verify that the services were performed. In addition, the written agreement is not sufficient. The agreement must be communicated to the construction firm's financial department. The financial department must also keep adequate records of the extra work and file them by contract and project so they may be deducted from the billing from the subcontractor.

The construction firm's field personnel may not want to risk disruption or delay in the project completion. Remember that it is the responsibility of the construction firm's field personnel to complete the project on time and within specifications. Field personnel may decide to perform the work requested rather than risk delay in completing the project. But it would be better if this were a conscious decision for the construction firm and the financial department and field supervision were aware of the decision. This is especially true if some of the construction firm's field personnel reported their time worked against the project number of the subcontractor rather than their own.

The construction firm always has the option of initiating legal action against the subcontractor to recover for the work performed. However, this alternative is expensive and proof of request and performance of the work may not be available. To justify the expense of legal action, the work performed for the subcontractor must have been extensive and material in amount.

EXTRA WORK PERFORMED FOR OWNERS

Although technically not back charges, extra work performed for owners at the job site during construction causes similar problems for the construction firm and therefore are classified the same way. Usually the owner's representative at the construction site requests the contractor's field personnel to perform a service or make a change that is not included in the contract. These may be the addition of a design, the addition of a color, a change in color, a change in location of a power source, the addition of a support, or the substitution of a more expensive material.

The request is usually verbal and the construction firm's field personnel, if the material or service is readily available, will perform the work rather than risk a delay or disruption in the completion of the construction project. Again, this is not a conscious decision of the construction organization. The construction firm's financial department is usually unaware of performance of the extra work and the possibility of reimbursement for the work performed is remote.

Again, if the construction firm's managers decide that it is necessary, as with back charges, the firm can initiate legal action against the owner or owners. This reaction will require proof of performance of the work and that the work was performed at the request of a representative of the owner who has the authority to bind the owner to such agreements. Legal action will usually disrupt or delay the construction project and will benefit only the attorneys. The courts should be used if the amounts are significant and justify the costs.

SUMMARY

Included in this chapter is a review of three related and important functions—the invoicing of the customer, the classification and control of the receivable from the customer, and the collection of payment from the customer. The discussion of invoicing emphasizes the need to conform to contractual provisions, the input from field personnel, and the controls required for invoice preparation. Also discussed are various types of invoices used in the construction industry such as progress billings, final billings, and the billing of retention. Additionally, the accounting treatment of invoices is presented.

Accounts receivable control and subsidiary accounts are described and illustrated. The problems of customer deductions, both authorized and unauthorized, customer retentions, and the application of payments to accounts receivable are detailed. The analysis includes a description and an illustration of the aging of accounts receivable.

The discussion of cash receipts emphasizes the need for control of the receipt of both cash and checks. Procedures to support control are recommended. Credit policies, which are closely related to the receipt of cash, are described. These policies are classified as preventive, corrective, or punitive.

The last part of the chapter is devoted to a subject peculiar to the construction industry; back charges. Back charges are described in detail as are the causes and possible solutions. These charges are the result of contacts between the construction firm's field personnel and representatives of other firms. It must be realized that these costs may never be recovered by the construction firm and should be treated as part of the costs of the construction project.

EXHIBIT 9. An Example of a Control Account

Accounts Receivable Control

Date	Explanation	Folio	Debit	Credit	Balance
Jan. 31	Balance				$150,123
Feb. 3		SJ 23	$134,192		284,315
Feb. 13		CR 10		$ 50,000	234,315
Feb. 15		CR 11		$104,315	130,000
Feb. 26		SJ 25	200,000		330,000
Feb. 28	Balance				$330,000

Accounts Receivable Subsidiary Ledger

Date	Customer	Balance
Feb. 28	Lee Construction Company	$ 55,000
Feb. 28	Walton Concrete Company	20,000
Feb. 28	Round and Boot Engineering (Project #23)	100,000
Feb. 28	Round and Boot Engineering (Project #29)	125,000
Feb. 28	Baltimore County Project #100346.	30,000
	Total accounts receivable by customer	$330,000

NOTES:

(1) The accounts receivable control account is included in the construction firm's general ledger.

(2) The subsidiary ledger, which supports the control account, is maintained by separate personnel and is located separately from the control account.

(3) The subsidiary ledger may be a separate ledger card for each customer, it may be a folder by customer of only unpaid or partially unpaid invoices, or it may be a listing of unpaid balances by customer in a computer.

(4) The customer listing of the subsidiary ledger is prepared at least monthly.

(5) The subsidiary ledger listing must equal the balance in the accounts receivable control account.

(6) If the subsidiary ledger and the control account do not equal, the reason for the difference must be identified and a correction made.

(7) The customers may be listed in the subsidiary ledger listing in the same order in which they are filed in the subsidiary ledger, in alphabetical order, or with the largest "balance due" customer first and then in descending order.

(8) A separate control account and subsidiary ledger may be used for each type of account receivable, such as progress billings, final billings, or retention invoices.

EXHIBIT 10. An Example of an Aging of Accounts Receivable

Description	Amount
Current balances	$1,200,000
Balances 1 to 30 days past due	400,000
Balances 31 to 60 days past due	200,000
Balances 61 to 90 days past due	110,000
Balances 91 to 120 days past due	80,000
Balances 121 to 150 days past due	50,000
Balances 151 to 180 days past due	40,000
Balances 181 plus days past due	30,000
Total accounts receivable balance	$2,110,000

NOTES:

(1) The accounts receivable aging can be prepared for the total accounts receivable or for each classification of accounts receivable such as receivables from progress billings, receivables from final billings, and receivables from retention.

(2) The days past due are calculated from the end of the terms of sale; thus terms of net 30 would not be past due until the 31st day from the date of sale. Invoices that require payment upon submission are past due from the date of submission until payment is received.

(3) Retention is not past due until the construction project is completed and accepted.

(4) The schedule of the aging of accounts receivable can include a column for the estimated loss percentage for each aging category and is usually historically based with modifications for current conditions.

(5) A column may also be added for the estimated loss from nonpayment by customers. This amount is the product of multiplying the estimated loss percentage for each category by the amount of receivables in each age group and summing these products.

(6) The estimated loss column is the basis for the adjustment of the allowance for doubtful accounts. The total of the estimated loss column is compared to the general ledger account "allowance for doubtful accounts." If the balance in the general ledger account is less than the total of the estimated loss column, the following journal entry must be made: debit the difference amount to bad debt expense; credit the difference amount to allowance for doubtful accounts. This entry will adjust the balance in the allowance for doubtful accounts to the total of the estimated loss column.

QUESTIONS

1. Of what importance to the construction firm are the contractual invoicing provisions?

2. What inputs are required for the preparation of invoices to customers?

3. How can invoice preparation be controlled to verify that all invoices, but only authorized ones, have been prepared?

4. What type of information is required on customer invoices?

5. Describe the various types of billings to customers.

6. Describe the accounting treatment, including the detail needed, for recording invoices to customers.

7. How many different kinds of accounts receivable may a construction firm have?

8. Are customer deductions important to the construction firm?

9. Why is retention a problem? What is the nature of retention?

10. What is cash application?

11. Prepare an example, and then describe the nature, of an accounts receivable aging.

12. How should cash received by the construction firm be controlled?

13. How should checks received in payment of accounts receivable be controlled?

14. What are lock boxes and how are they used?

15. Does the construction industry have cash flow problems? If so, describe these problems.

16. Describe and classify the various types of credit policies that a construction firm might use.

17. What are back charges?

18. How can the disputes that accompany back charges be eliminated?

19. Are back charges an ordinary and necessary expense of a construction project? Explain.

REFERENCES

Anthony, Robert N., and James S. Reece, *Management Accounting* (Homewood, Illinois: Richard D. Irwin, 1975).

Chorba, George J., *Accounting for Managers* (New York: American Management Association Extension Institute, 1978).

Dearden, John, *Cost Accounting and Financial Control Systems* (Reading, Massachusetts: Addison-Wesley Publishing Co., 1973).

Gordon, Myron J., and Gordon Shillinglaw, *Accounting: A Management Approach* (Homewood, Illinois: Richard D. Irwin, 1974).

Horngren, Charles T., *Cost Accounting: A Managerial Emphasis* (Englewood Cliffs, New Jersey: Prentice-Hall, 1977).

Kieso, Donald E., and Jerry J. Weygnadt, *Intermediate Accounting* (New York: John Wiley & Sons, 1977).

Meigs, Walter B., A. N. Mosich, and E. John Larson, *Modern Advanced Accounting* (New York: McGraw-Hill Book Company, 1979).

Montgomery, A. Thompson, *Managerial Accounting Information* (Menlo Park, California: Addison-Wesley Publishing Company, 1979).

"Professional Notes," *The Journal of Accountancy*, December 1979.

Pyle, William W., John Arch White, and Kermit D. Larson, *Fundamental Accounting Principles* (Homewood, Illinois: Richard D. Irwin, 1978).

Rayburn, L. Gayle, *Principles of Cost Accounting with Managerial Applications* (Homewood, Illinois: Richard D. Irwin, 1979).

Welsch, Glenn A., Charles T. Zlatkovich, and Walter T. Harrison, Jr., *Intermediate Accounting* (Homewood, Illinois: Richard D. Irwin, 1979).

CHAPTER 7

OPERATIONAL ASSETS AND TAXES

The control and use of capital operational assets are related and are key functions in the construction industry. For a construction firm to be profitable, it must use its operational assets intensively. To use its operational assets intensively, the firm must know where they are located and have a plan for their future use.

Taxes are the burden of all business and individuals, as well as of construction firms. However, in the construction firm a modification of the tax law can be disastrous if the modification was not anticipated when the contract was written.

THE ACQUISITION OF OPERATIONAL ASSETS

The acquisition of operational assets by business firms must meet the accounting requirements for such acquisitions. The first requirement is the definition of an operational asset. An operational asset is an asset that will be used in the conduct of the business and whose use will affect the firm for more than one fiscal period. That is, the operational asset will produce revenue for more than one accounting period (fiscal year).

If an item meets the definition of a capital operational asset, the firm must decide at what value to capitalize the asset. The general rule is that the value of a capital operational asset is the purchase cost plus all costs necessary to put the asset in place for use by the purchaser. These costs include, but are not limited to, installation, transportation, testing, structure modification (related to that asset), and training costs for the purchaser's employees.

In addition, and within the accounting rules, most firms have policies defining the nature of an operational asset. The most commonly used policies define an operational asset as having a life in excess of one year and costing at least a minimum dollar amount, such as $5,000. The dollar minimum is applied because of the accounting principle of materiality. Although the accounting rules concerning operational assets are silent about dollar amounts, it would not be cost efficient to treat items of small dollar value with the same care as capital operational assets. The principle of materiality pervades all accounting rules and principles.[1]

EVALUATING THE PURCHASE OF AN OPERATIONAL ASSET

The identification of a need for an operational asset should start a process of evaluation, discussion, and reflection. The commitment of the future is a serious decision. An operational asset can be purchased or leased with or without the option to buy at the expiration of the lease. With a purchase the firm accepts all the usual responsibilities of ownership. The firm is responsible for maintenance, insurance, and damage to the asset. Leasing usually includes an option to purchase the asset at the termination of the lease and includes all the responsibilities of ownership during the lease period as well as after the asset has been purchased.

The managers of the construction firm must decide whether to purchase or lease the equipment or asset. There are two aspects to the decision. One is the financial aspect: will it cost less to purchase or lease the asset? Two, what are the nonquantitative reasons for purchasing the asset, if any. The accounting and financial division of the firm will prepare an analysis of the financial impact. The purchase alternative, including the asset cost, the interest on the purchase loan, the depreciation tax shield, and the applicable taxes, is compared with the lease costs. The lease cost includes the monthly lease payment plus any taxes to be paid by the lessee reduced by any reassignment of the investment tax credit from the lessor to the lessee. Unless the lessor, who is usually a financial lessor and does not want to own the equipment, can justify a faster writeoff of the asset, the two alternatives will cost about the same. In accounting terms, the difference will not be material. Therefore, the ultimate choice of an alternative will be based on nonquantitative data, such as the desire to own all assets or this particular asset. The firm's

[1]Donald E. Kieso, and Jerry J. Weygandt, *Intermediate Accounting* (New York: John Wiley & Sons, 1980), pp. 473-496.

managers may decide that they do not want assets that are not owned by the firm on the premises.

Once the analysis has been completed and the alternative chosen, the decision must be formally approved and recorded. The record can be in letter form, but it should have the signatures of the financial division and of general management. Execution of the decision should not begin until the formal decision has been received and recorded.

THE PHYSICAL CONTROL OF OPERATIONAL ASSETS

Again, the wide geographic dispersion of a construction firm's operational assets complicates the physical control process. The firm's managers should always be able to identify the number of operational assets the firm has and their locations. In a large construction firm there will always be some operational assets that are in transit. However, that information should be available to the firm's managers.

One method for identifying operational assets is to affix a plate or metal tag to the asset. The tag would include the firm's name and a sequential number that would be that asset's tag number. A sequential check would identify missing assets. Because data processing equipment can process numerical data less expensively, the tag number is the basis for identifying and locating operational assets. All assigned numbers must be accounted for. Files can be kept by number as well as by asset description and each can be cross-referenced to the other.

Operational asset location files should equal the operational asset accounts in the general ledger. A method for developing location files must be compatible with the operations of each construction firm. One approach is to list all of the firm's operating locations, adding new projects as they are started and dropping completed construction projects. Each location on the list is assigned a number, and the location number is posted to the file of each asset within that location. Field personnel must notify the operational asset control group when an operational asset is moved. The construction firm's managers must know where each operational asset is located to use it efficiently.

To be profitable, a construction firm must use its operational assets intensively. Each asset, if optimum results are achieved, should be used every hour of its life. It is not possible to use assets effectively without planning and scheduling their use. A requirement is knowledge of the asset's present location and a plan for its future use. The objective of scheduling the use of operational assets is to charge each construction project with the hours and cost of use of every operational asset. To meet this objective, accurate reporting of the use of operational assets by field personnel is a necessity.

DEPRECIATION POLICY

Depreciation is the writing off of a portion of the construction firm's assets (except land) during each fiscal year. The journal entry is: debit: depreciation expense—equipment (or machinery, or buildings); credit: accumulated depreciation—equipment (or machinery, or buildings). The expense account reduces earnings, is a nominal account, appears in the income statement, and is closed at the end of the fiscal year. Thus it is a nonfund or noncash item which is part of operating costs. The amount subject to depreciation is the acquisition cost (as described above) less any residual or scrap value.

DEPRECIATION METHODS

There are a number of depreciation methods that may be applied to the amounts subject to depreciation. One is the unit of production method. For the construction firm this would be the recognition of depreciation for each hour the machinery or equipment is used. The estimated life of the machinery, expressed in total hours of life, is divided into the depreciable amount to compute a rate of depreciation per hour of use. This is the more useful method of depreciation for the construction firm because it is compatible with the method used for charging the use of equipment to construction projects (to be discussed below). However, the problem is the unanswered question of the depreciation of the machinery and equipment even if it is not in use. One technique for overcoming this weakness is to reduce the estimated total hours of life to recover in the hourly charge for periods when the equipment is idle.

Another method is the straight line method. The straight line method is time oriented. The life of the equipment (or machinery) is estimated in total years. The years are then divided into the depreciable amount and a depreciation rate per year is the result. This rate is used to record the decrease in book value of the machinery and the depreciation expense. The weakness inherent in the straight line method is that it may not measure the true depreciation of the machinery or equipment. The straight line method's underlying assumption is that the depreciation of machinery and equipment (or buildings) is time based. This may not be true, depending on the nature of the machinery and equipment and the intensity of its use. However, most firms, including construction firms, use the straight line method for financial statement purposes and another method (to be discussed below) for income tax return preparation. The straight line method is used for financial statements because it normally reduces revenue less than the other methods and the firm can, compared to other depreciation methods, report a higher net income.

A popular accelerated method of depreciation is the sum of the years digits. The accelerated methods, including the sum of the years digits, are used more widely for the preparation of income tax returns than in the financial statements. The disadvantage of the sum of the years digits method (and all accelerated methods) for financial statement purposes is that they result in a lower reported net income than the straight line method.

The sum of the years digits method of depreciation is calculated as follows. Assuming a five year life for the equipment, the years of life are summed: year $1 + 2 + 3 + 4 + 5 = 15$, the first year's depreciation is equal to 5/15 of the acquisition cost (less salvage value) of the equipment, the second year's depreciation is equal to 4/15 of the acquisition cost. For the remaining years of life of the equipment, the same technique is applied. This method is popular for income tax purposes because it causes larger depreciation deductions in the early years of life of the equipment. This reduces the outflow of cash from the firm to governments. The firm has the use of cash now to invest and earn a return, which will further increase their cash inflow.

Perhaps the most widely used accelerated depreciation method is the declining balance method. Technically, its name is double the straight line rate applied to the declining asset balance method. Because it is an accelerated method it has the same disadvantage as other accelerated methods when used for financial statement purposes. It causes a lower net income to be reported on the financial statements than the straight line method.

The double declining balance method is calculated as follows. Assuming a machine with a 10 year life, the straight line rate is 10%. This rate is doubled and applied to the acquisition cost of the asset for the first year to determine the depreciation expense. For the double declining balance method the acquisition cost is the amount before the deduction of any residual or scrap value. The depreciation for the second year is calculated by applying 20% (double the straight line rate) to the acquisition cost less the first year's depreciation. The third year's depreciation equals 20% times the acquisition cost less the first year's and the second year's depreciation expense (this is the declining balance). Like the sum of the years digits method, the declining balance method is used for income tax return preparation rather than preparation of financial statements. It is used for income tax returns because it preserves the firm's cash flow.

Composite depreciation methods can be used with any of the depreciation calculations discussed above. For composite use, however, the depreciation expenses are reduced to a rate for each item which is averaged and applied to the balance in the composite asset account. The composite asset account is usually used for recording and grouping assets of small value. It would not be used for cranes, front end loaders, or eighteen wheelers. It would (or could) be used for drills, compressors, and small mixers. The composite operational asset account can be of two kinds: (*a*) include only assets that are alike, that is homogeneous or (*b*) include all assets of a certain value range, regardless of nature, in one asset account.

For a homogeneous composite operational asset account the rate of depreciation used for that type of asset is applied to the balance in the asset account. This amount (the balance times the rate) is debited to the depreciation expense account and credited to the accumulated depreciation. For a composite asset account that includes unlike items, a study is made of the type of assets in the account, and the amount of depreciation for each is calculated. The asset acquisition costs and the depreciation for each are totaled. The total depreciation for the year is divided by the total acquisition cost and a rate is established. This rate is used to calculate the depreciation expense for this account. With an account that includes unlike items, a study should be done of the assets in the account at least every six months to verify the accuracy of the depreciation rate.[2]

Usually, with composite operational asset accounts, the depreciation calculated for the financial statements and for the tax returns is the same. In addition, regardless of which of the depreciation methods described above is used for the financial statements, the financial statements must include the acquisition cost of the assets less the accumulated depreciation through the statement date, such as: Machinery $100,000 (less $40,000 of accumulated depreciation as of the statement date) equals net book value of machinery $60,000. The statement would look as follows:

[2]*Ibid*., pp. 519-540.

Operational Assets:

Machinery	$100,000	
Less: accumulated depreciation	(40,000)	$60,000

THE PERIODS USED FOR RECORDING DEPRECIATION

There are a number of policies for determining when depreciation will be calculated, regardless of the method chosen or the lives used. The choice of period is usually a function of the cost to compute depreciation for partial months or periods and whether the effect of excluding the amounts in question is material, that is, whether the exclusion will have a material effect on the interpretation of the financial statements. For the construction firm the criteria are different. Because, as discussed below, the firm must recover the depreciation of the machinery and equipment through hourly use charges to construction projects, the construction firm should compute partial depreciation for fractions of months.

Business firms that calculate depreciation on a monthly basis can use a number of variations for recording the expiration of the value of its operational assets. One policy is the recording of depreciation in full monthly increments only. If a machine is purchased (received, put into use) within a month, no depreciation is recorded until the first full month of use and then a month's depreciation is recorded. When this equipment is disposed of, a full month's depreciation will be recorded for any partial month of use before disposal.

Another policy is to compute partial depreciation but not for less than one half a month. Any equipment used for less than half a month at purchase or disposal is charged for one half a month of depreciation. A less popular method of computing depreciation is to calculate the depreciation daily based upon a 360 or 365 day year.

One widely used policy for computing depreciation is to record six months' (half year's) depreciation in the year of purchase, regardless of when purchased within the year, and record six months' (half year's) depreciation in the year of disposal regardless of when the asset is sold within the year. The purpose of this policy is to reduce the clerical costs of recording and maintaining records of depreciation. In most cases, this policy will not have a material effect on the interpretation of the firm's financial statements.

The most popular depreciation policy is the recording of depreciation in yearly increments. There are at least two variations to this policy. One variation is to record depreciation only after the first year of purchase and to record a full year's depreciation in the year the asset is disposed of regardless of when the asset is disposed of within the year. The other variation is to record a full year's depreciation in the year of purchase regardless of when purchased within the year, and to record no depreciation in the year of disposal regardless of when the asset is disposed of within the year.

Either of these variations may be used but as with any of the policies described above, the policy or variation chosen must be consistently applied. The purpose of the full year's depreciation policy is to simplify and reduce the costs of recording and maintaining records of depreciation. Again, this policy will not materially misstate the firm's financial statements.

The reader must realize that the method of depreciation chosen for use with the financial statements is not required to be used with the income tax returns and vice versa. The object of our current income tax laws and regulations is to allow the firm the largest and earliest tax deduction possible to preserve cash flow while allowing the firm to show the maximum net income possible to its shareholders on the financial statements. In practice, depreciation policy is usually consistently applied from year to year. While depreciation methods are not consistently applied between the financial statements and the income tax returns, they usually are consistently applied within each of these areas.[3]

The two broad criteria for deciding which depreciation method to use for income taxes and which for the financial statements are: (a) for the financial statements choose the method that will result in the largest reported net income and for the income tax return choose the method that will result in the lowest tax liability; or (b) choose the depreciation method for the financial statements that will cause the depreciation cost to reflect the most accurate possible expiration of the value of the operational assets and for the income tax return choose the method that will result in the lowest tax liability.

CHARGING THE COSTS OF OPERATIONAL ASSETS TO CONSTRUCTION PRODUCTS

The construction firm must be able to charge the costs of its operational assets to the construction projects on which they are used. The costs that are charged to each project must include the depreciation costs, the costs of operators, fuel costs, repairs, insurance, property taxes, and any additional depreciation related to

[3]*Ibid.*, pp. 541-555.

major overhauls or betterments. Because there is a lag in the recording of these costs the construction projects must be charged at an estimated rate. However, unless the rate is specified in the contract, the rate should recover all of these costs. If possible, it would be more appropriate for the rate to exceed these costs and result in an additional profit to the construction firm.

Charging these costs to construction projects is essential not only from the perspective of cost recovery, but also for the need to estimate project charges in the future. All of the costs associated with operational assets should be accumulated in their own accounts and then transferred to the clearing account used for the charge to the construction project. For example, the journal entry for recording depreciation would be as follows:

	Debit	Credit
Depreciation expense—equipment	$XXXXXX	
Accumulated depreciation—equipment		$XXXXXX

The journal entry for the use of the equipment would be based on the number of hours the equipment was used on the project times a rate for that type of equipment and would take the following form:

	Debit	Credit
Construction project #XXXXXX— equipment use	$XXXXXX	
Equipment charged to projects clearing account		$XXXXXX

At the end of the accounting period, the depreciation and other costs would be closed to the equipment charged to projects clearing account as follows:

	Debit	Credit
Equipment charged to projects clearing account	$XXXXXX	
Depreciation expense—equipment		$XXXXXX

All costs of operational assets used for construction would be journalized as outlined above regardless of whether the asset was included in a composite account or recorded separately. Usually, the equipment charged to projects clearing account will have a balance after all of the above entries have been made. No one is able to estimate accurately enough to equate the debits and credits.

If the remaining balance in the clearing account is a debit, the actual costs of operating the asset exceeds the estimate for charging projects and the projects have been undercharged. An entry must be made crediting the clearing account and debiting the asset charge account in each construction project. The additional charge should be allocated on the basis of assets costs already charged to each project. The effect of the entry is to increase the cost of the use of assets for each construction project.

Should the remaining balance in the clearing account be a credit, each project has been overcharged for the use of equipment. The adjusting entry is a debit to the equipment charged to projects clearing account and a credit to the equipment use accounts in each construction project. The result of the adjusting entry is to reduce the costs of the use of equipment for each construction project.

Another acceptable approach is to charge the difference in the equipment charged to projects clearing account to either miscellaneous income or expense accounts. If the balance in the clearing account is a debit, the actual costs are greater than the charges to projects and should be debited through an adjusting journal entry to a miscellaneous expense account. If the balance in the clearing account is a credit, the actual costs are less than the estimated charges to projects and the clearing account must be debited and a miscellaneous income account credited. The only serious disadvantage of this approach is that the equipment use accounts for each construction project will be understated. However, if the balances in the clearing account are not material in amount, this approach will suffice and not cause any significant inaccuracies in the financial statements.

For the recording of equipment use by project to be successful, field personnel must regularly report the use of operational assets for each construction project. The report should include the actual hours of use for each major item of equipment and an estimate for the hours of use for assets of lesser value. The objective of charging the equipment costs to projects are to recover the costs of equipment use through project revenue and to provide a reliable history of the cost of construction projects for use in bidding on work in the future.

INSURANCE

In the construction industry, as in other industries, the firm must provide all the necessary insurance coverage. Liability (personal) insurance is necessary for protecting the firm from legal action in case of accidents and injury on the firm's property to employees or visitors. Medical insurance is usually provided for all employees. In some cases the insurance is contributory on the part of the employees or paid for totally by the employer. Usually for salaried employees the firm administers the medical insurance through an insurer. For union employees the medical insurance program may either be administered by the firm or through a union administered benefit fund.

In most states workperson's compensation insurance is required. The insurance is placed with an insurance carrier and the rates are based on experience. The rates charged the firm for the insurance are computed on the amounts and types of the firm's payroll including factors for the inherent risk of the various job classifications. For the construction firm the costs of workperson's compensation are relatively high.

Construction firms typically use large numbers of various types of motor vehicles. Therefore, the firm must have all the usual coverage on the vehicle itself in case of loss, liability insurance in case of loss to others from its use, and insurance on the occupants of the vehicle in case of injury.

The costs of insurance are part of the cost of operational assets and are included in the hourly charge to construction projects. These costs are in addition to the operators' wage and fringe benefit costs. The costs are also included in the adjusting journal entry debit to the equipment charged to projects clearing account.

INSURANCE JOURNAL ENTRIES

There are two alternative methods for recording insurance costs for the fiscal period. These entries can be used with any of the types of insurance required. The possible exception is workperson's compensation insurance for which, depending on the experience rating, a refund may be due at the fiscal year end and may have to be estimated and recorded.

One alternative for recording insurance costs is to debit insurance expense when the premium liability is recognized and credit a liability such as vouchers payable. At the end of the fiscal year any of the insurance premium that is not used or applies to future years is recorded as an asset. To record the unexpired insurance at fiscal year end, the unexpired insurance account is debited for the amount which applies to future years and the insurance expense account is credited because that is where the total premium was originally charged.

The other alternative is to record the insurance costs (debit) in unexpired insurance. This account is a real account and an asset account usually classified as a current asset; however, some accountants classify it as an other asset. At the end of the fiscal year the costs of the insurance that have been used, that is, the insurance costs of coverage that has expired, is debited to the insurance expense account and credited to the unexpired insurance account. With both methods the balance in the unexpired insurance account is the same. The balance in the insurance expense account applicable to equipment is transferred to the equipment charged to projects clearing account.

Insurance is an example of the cash flow difficulties the construction firm must resolve. Almost without exception, insurers require payment in advance. The construction firm will be reimbursed through billings on its various construction projects. However, reimbursement may take months. Meanwhile, the construction firm must pay the insurer and find a method for financing the payments until reimbursement is received. In most cases the construction firm must notify its insurers of construction projects which are completed and of projects that are started. This requires that the construction firm provide the resources necessary to check the accuracy of the charges from the insurer.

Although payment and performance bonds are usually purchased through insurance companies, they are not included in this section. Because they are a special kind of insurance, they are discussed in a later section.

PROPERTY TAXES

In some states there are two types of property taxes: personal and real. However, even in those states that have both personal and real property tax laws the personal property tax provisions of the law may be inactive.

PERSONAL PROPERTY TAXES

The typical construction firm has a large number of assets that qualify as personal property and, therefore, are subject to tax. Examples of personal property are inventory, front end loaders, dump trucks, mixer trucks, and office furniture. If the jurisdiction in which the construction firm has projects imposes a personal property tax, the firm must keep a list of all personal property within that jurisdiction which is subject to tax. In addition, a value must be assigned to each item. The firm's value and the calculation of the tax should be checked against the tax invoice submitted by the jurisdiction.

A major problem is valuing personal property subject to tax. In some cases, book value is available; in other cases, there is a ready used asset market. The taxing jurisdictions usually attempt to base the tax on the market value or the approximate market value of the asset. For the firm, the establishment of a market value for many of the assets that are subject to personal property tax would be more costly than the tax! Under these circumstances construction firm personnel should review the values included in the jurisdiction's tax invoice and only investigate those values that appear to be extreme.

REAL PROPERTY TAXES

There are relatively few jurisdictions that do not impose a real property tax. Again, the construction firm must maintain a complete list of all of its assets that are real property and must include in this list the jurisdictional location. Because of the high value of real property, the construction firm must establish and revise the tax value (which is usually the market value) of its real property. The tax value ordinarily includes the estimated market value of the property and the tax rate varies depending on the use classification of the property. Therefore, as a base for a tax appeal and to protect cash flow, the construction firm must establish its own values and monitor the use classification by jurisdiction of all of its operational assets that are real property and subject to tax.

Real property taxes must be paid when due. If these taxes are not paid when due the firm can be subject to heavy penalties and in some cases the property may be liened. To keep the title to its real property free of liens and to avoid the penalties, the construction firm must meet the obligations for its real property taxes despite any delay or lag in reimbursement for the use of the real property for a construction project.

The recording of real and personal property taxes is not complicated except in the timing of the entries. Real and personal property taxes should be recorded as an accrual, regardless of when paid, over the fiscal year of the taxing authority or jurisdiction. For the entries listed below it is assumed that the taxes on real and personal property are recorded as expenses. For financial statement presentation real property taxes imposed on land that is held for sale (inventory) are capitalized as part of the cost of the land inventory to be matched with the revenue from the sale of the land. The real property tax on land used as an asset in the business is recorded as an expense.

The journal entries for recording real and personal property taxes and the transfer of the costs of personal property to construction projects follow. For recording the personal property tax: debit: personal property taxes expense; credit: vouchers payable. For transfer of the tax to construction projects: debit: equipment charged to projects clearing account; credit: personal property taxes expense. For recording the real property tax: debit: real property taxes expense; credit: vouchers payable.

INCOME AND OTHER TAXES

Income taxes are not limited to the federal government. Many states and subdivisions (counties and cities) also have income tax laws, some of which, such as New York City's, are very complex. In addition, there are jurisdictions that do not enforce or do not have a personal property tax but impose a special inventory tax. Sales and use taxes are almost universal. For the construction firm with its diverse geographic locations, the minimizing of the income tax paid is a complex responsibility. Therefore, this book concentrates on the recording rather than the calculation of the income tax liability.

FEDERAL INCOME TAXES

The object of the preparation of the income tax return for the construction firm is to minimize the federal income tax liability. The return should be prepared by specialists either within or outside the firm. Usually the outside specialists are the firm's independent CPAs. The preparation of the federal income tax return requires a professional specialist.

The federal income taxes are recorded as an expense with a debit to federal income tax expense and a credit to federal income taxes payable. As was discussed earlier in this book, the federal income tax expense must be classified with the income that caused it within the income statement. Federal income tax expense associated with income from operations must be deducted from that income; federal income tax from extraordinary items must be deducted from those items in the income statement. The payment of the income tax liability is debited to federal income tax payable and credited to cash.

STATE INCOME TAXES

As with federal income taxes, state income tax returns must be prepared by professional specialists. These specialists may be employees of the construction firm or outside independent CPAs. The major problem for the construction firm with state income taxes is that of double taxation. Most state income tax laws provide for credits for taxes paid to another state. However, it is not always clear who has primary taxing jurisdiction. The resolution of these disputes requires knowledgeable professionals and sometimes attorneys.

The state income taxes are recorded as an expense with a debit to state income tax expense and a credit to state income taxes payable. State income taxes associated with income from continuing operations must be deducted from that income. State income tax caused by operating income or from extraordinary items must be deducted from these items in the proper section of the income statement. The payment of the state income tax liability is recorded by a debit to state income taxes payable and a credit entry to cash.

SUBDIVISION INCOME TAXES

Subdivision income tax returns must also be prepared by professionals. Again, the construction firm has a choice of whether to employ its own specialists or hire outside specialists. The difficulty with subdivision income taxes is jurisdiction. Such questions as who should be subject to the tax, how much of a firm's income should be taxed, and what credits are granted for payment of income taxes in other jurisdictions need answers. These questions may arise whether the subdivision income tax involves complex calculations, as New York City's, or is simple to calculate, such as Maryland's piggyback tax which includes a statute determined percent of the state tax paid to the county and collected by the state from taxpayers.

Subdivision income taxes are recorded as an expense with a debit to subdivision income taxes expense and a credit to subdivision income taxes payable. Subdivision income taxes associated with income from continuing operations must be deducted from that income. Subdivision income tax caused by operating income or from extraordinary items must be deducted from these items in the proper section of the income statement. The payment of the subdivision income tax is recorded by a debit to subdivision income taxes payable and a credit to cash.

INVENTORY TAXES

Some jurisdictions impose an inventory tax on the value of inventories located within their boundaries. The tax is calculated as a percent of the inventory value (a mill rate). The methods of verifying the inventory value vary and the reporting requirements range from the reporting of the year end book value of the inventory to the average inventory book value for the tax year. In some instances the market, rather than book value, must be used. Some jurisdictions audit the information reported, others rely on the taxpayer's information.

If the inventory tax cannot be avoided by locating the inventory in a nontax jurisdiction, the following journal entries are required to record the tax liability and subsequent payment. To record the tax liability: debit: inventory tax expense; credit: vouchers payable—inventory tax. To record the payment of the tax: debit: vouchers payable—inventory tax; credit: cash.[4]

SALES AND USE TAXES

The probability is that the construction firm will have projects within states with sales and use tax statutes. The application and interpretation of these laws relative to construction are more complex than in the manufacturing environment. The construction firm's managers must keep detailed and accurate records of the purchase and use of all materials. The "use" provisions of most sales and use tax laws provide for taxing materials used in a state regardless of where purchased. Most of these laws provide for credit for sales and use taxes paid to other states but only to the extent that the payment was equal to the sales tax payable in the location of use.

In addition, the tax burden is applied to both the purchaser and the seller. If the jurisdiction cannot collect the tax from the buyer it can collect from the vendor. In one case, the construction firm claimed exemption from the sales tax for the application of asphalt to a highway through an asphalt spreader. The exemption was based on the argument that the spreader was a manufacturing asset used to apply asphalt to the roadway. The exemption claim was rejected and the construction firm was forced to absorb the sales tax cost for that project.

The construction firm must include in its construction contracts provisions for changes in the interpretation and application of sales and use tax statutes. These contractural provisions must permit renegotiation of the sales and use tax burden if there is a change in the interpretation or application of the law. The effect on construction contracts of a change in the rate of sales and use tax imposed is also unclear. Usually the revision in the rate (upward is the most likely change) is delayed to permit the conclusion of existing purchases and contracts. However, there is doubt concerning how long the imposition of the new tax rate can be delayed. For long term construction contracts the time allowed may be sufficient to readjust

[4]*Ibid.*, pp. 880-910.

the contract for the increase but not to complete it. If this is the interpretation of the law, the construction firm must include re-negotiation provisions for sales and use taxes in its construction contracts.

The sales and use tax collections should be paid to the taxing authority within the prescribed payment period. If, for example, as within some jurisdictions, the sales and use tax collected is due and payable on the twentieth day after collection, the construction firm should make every effort to meet that payment date even if the exact amount due is not available. The firm should pay an estimated amount if the actual amount is not known. By law, penalties and interest can be collected for late payment. It is possible for the taxing authority to invoke the penalty for payments that are one or more days late. Interest is collectible for each day the payment is past due. However, penalties and interest are seldom enforced if the payment is made within a reasonable number of days from the due date.

Interest is usually charged at the prime rate for the location. But the penalty invoked is almost always severe. For example, in 1977 the state of Maryland imposed a penalty for late payment of 25% of the sales and use tax due. Although the firm must attempt to avoid tax penalties, if a penalty is invoked an appeal should be filed. In each jurisdiction there is a procedure for appealing interest and penalties. Often the appeal is nothing more than a telephone call. It is rare for a jurisdiction to force a firm into financial difficulty because of late tax payments and lose the employment the firm provides for the jurisdiction's citizens.

The journal entries to record the liability for sales taxes and the subsequent payment are not difficult. The debit entry is to accounts receivable unless the construction firm cannot recover the tax from the customer. Then the debit entry is to Sales Tax Expense with a project designation. The credit entry is to sales and use tax payable. The payment is recorded by a debit to sales and use tax payable and a credit to cash. If a penalty and interest charge has been enforced a debit entry is required to interest and penalty expense—sales and use taxes and a credit entry to interest and penalty payable. Also, the debit entry should be detailed by construction project. The payment is recorded by a debit to interest and penalty payable and a credit to the cash account.

SUMMARY

This chapter concentrates on operational assets and taxes. The accounting rules that guide the acquisition of operational assets are explained. The procedures for acquiring operational assets are described along with the need for financial evaluation of the alternatives. Requiring approvals for the purchase of operational assets is discussed as are the physical control and scheduling of the use of operational assets.

The various depreciation methods and policies are listed and defined. Included in these methods are the unit of production method, the straight line method, the sum of the years digits method, the double the straight line rate on the declining asset balance method, and composite methods. The depreciation policies include the monthly recording of depreciation, recording depreciation in six months' increments, and the yearly recording of depreciation. The use of different methods of depreciation for financial statement preparation that differ from those used for income tax return preparation is explained.

The need to charge the total costs of the use of operational assets to construction projects is explored. The total operating costs of assets are defined to include the costs of operators, including their fringe benefits, the depreciation of the assets, fuel costs, if any, costs of repairs, insurance costs, taxes, and any supplemental depreciation caused by major overhauls or improvements. The need for usage reports from field personnel is stressed. The accounting procedures for recording and evaluating the extent of use of assets for construction projects is explained.

The types of insurance needed by a construction firm are described, such as liability insurance, medical insurance, workperson's compensation insurance, and the various types of motor vehicle insurance. The accounting journal entries required to record the costs of insurance are given.

Real and personal property taxes are examined in this chapter. Valuation problems, the large numbers of items, assessment rates, tax classifications, and tax payment are discussed. In addition, the accounting recording and accounting treatment of property taxes are presented.

In the last section of the chapter, income and other taxes are probed. Emphasis is placed on the recording of such taxes as federal income taxes, state income taxes, subdivision income taxes, and inventory taxes. Sales and use taxes are included in a separate discussion. The difficulties of applying sales and use tax statutes to construction are examined. The need for contractural protection for the construction firm from changes in the sales and use tax statutes or their interpretation is noted. The need for timely payment of sales and use tax liabilities is emphasized. The chapter ends with an explanation of the accounting journal entries required to record the liability for and payment of sales and use taxes.

REFERENCES

Anthony, Robert N., and James S. Reece, *Management Accounting* (Homewood, Illinois: Richard D. Irwin, 1975).

Chorba, George J., *Accounting for Managers* (New York: American Management Association Extension Institute, 1978).

Dearden, John, *Cost Accounting and Financial Control Systems* (Reading, Massachusetts: Addison-Wesley Publishing Co., 1973).

Gordon, Myron J., and Gordon Shillinglaw, *Accounting: A Management Approach* (Homewood, Illinois: Richard D. Irwin, 1974).

Horngren, Charles T., *Cost Accounting: A Managerial Emphasis* (Englewood Cliffs, New Jersey: Prentice-Hall, 1977).

Kieso, Donald E., and Jerry J. Weygandt, *Intermediate Accounting* (New York: John Wiley & Sons, 1977).

Kieso, Donald E., and Jerry J. Weygandt, *Intermediate Accounting* (New York: John Wiley & Sons, 1980).

Meigs, Walter B., A. N. Mosich, and E. John Larson, *Modern Advanced Accounting* (New York: McGraw-Hill Book Company, 1979).

Montgomery, A. Thompson, *Managerial Accounting Information* (Menlo Park, California: Addison-Wesley Publishing Company, 1979).

"Professional Notes," *The Journal of Accountancy*, December 1979.

Pyle, William W., John Arch White, and Kermit D. Larson, *Fundamental Accounting Principles* (Homewood, Illinois: Richard D. Irwin, 1978).

Rayburn, L. Gayle, *Principles of Cost Accounting with Managerial Applications* (Homewood, Illinois: Richard D. Irwin, 1979).

Welsch, Glenn A., Charles T. Zlatkovich, and Walter T. Harrison, Jr., *Intermediate Accounting* (Homewood, Illinois: Richard D. Irwin, 1979).

Questions for Chapter 7 appear on page 208.

CHAPTER 8

PERSONNEL POLICIES, PROJECT DIRECT AND INDIRECT LABOR, AND INDIRECT CONSTRUCTION COSTS

Personnel policies of construction firms differ markedly from those of other industries. The nature of construction and its policies affect the ability of the firm to identify its project labor costs in a conventional manner. However, for its survival, the construction firm must identify direct and indirect labor and other indirect costs to construction projects. This assignment of costs is necessary to follow the progress of each project and to measure each project's final profitability. In addition, historical information is needed for the preparation of bids for future contracts. Regardless of the inadequacies of the available methods, the construction firm must attempt to allocate its labor and other costs to its construction projects.

PERSONNEL POLICIES

One circumstance that directly influences the personnel policies of the construction firm is the variety of unions and union locals with which it must deal. For example, one construction firm may have to negotiate with the operating engineers, the international association of machinists, various paving councils, the carpenter's union, and the teamsters. Because the contract expiration dates are not uniform the construction firm may be engaged in almost continuous negotiations. This absorbs some of the firm's resources which, therefore, are not available for other uses. Each of the various contracts will include differing provisions, such as the total payment required to employees (union members) for working only a partial day.

Some of the union contracts will require payment to the employee for seven hours, some for four hours, and others for the whole day regardless of the number of hours worked. In addition, each union may administer its own benefit fund. This will force the construction firm to pay contributions at different rates and at different times to each union administered benefit fund. Thus the result is an increase in the construction firm's administrative costs.

The nature of the industry also complicates the construction firm's ability to conform to equal employment opportunity statutes. Construction firms must comply with the equal employment opportunity statutes of the various states and the federal government. In addition, the firm must comply with any special equal employment statutes enacted by the subdivisions of states and any special provisions of its construction contracts. Governments often reinforce the equal employment statutes with specific contractual provisions. If the construction firm does not comply with the law, it is in violation of its responsibilitiess under specific contracts. And the violations may not be completely under the construction firm's control. It is accepted practice within the industry for the unions to furnish certain types of employees.

HIRING CONSTRUCTION WORKERS

Construction firms use both orthodox and unorthodox methods for hiring employees. They have employment applications and files, as in any other firm. Usually different policies and procedures apply for hiring salaried than for hiring hourly employees. Salaried employees are hired and retained under policies similar to those found in manufacturing. Although these or similar policies are applied to some hourly employees, most are hired and retained from a number of sources and through a variety of policies.

There are hourly employees who are considered by themselves and by the firm as being permanent (that is, under most circumstances) employees. These employees return season after season and are employees, full time, of one construction firm. The firm may also keep them busy during the off-season by making use of them in the maintenance shops or in the office. Regardless of how they are originally hired, they are almost continuously employed by one firm.

Often the foreman has a crew that he has kept with him as he has moved from firm to firm and from job to job. When he is hired by the construction firm it is with the understanding that he will bring his crew with him. In some instances crew members will move together because they have a long and successful history of working together. They may be kept together by an informal crew leader. The informal leader may arrange for the transfer or hire of the crew. He or she may also arrange that the foreman or supervisor with whom the crew works best be put in charge.

One source of a large number of hourly employees is the union hiring hall. Because a construction firm has projects at many locations, it is not economical or practical to transfer all the hourly employees needed from location to location. Therefore, the firm hires employees from the local union hiring halls. The firm calls or sends an order to the union hall for the number, type, and skills required. The union fills the order from members who are awaiting work either at the hall or in a nearby location.

Although an economical and practical alternative is not available, the use of union hiring halls can adversely affect the relationships between employer and employee. For example, the use of the hiring hall can lead to a lack of continuity in employment by the construction firm. The firm usually accepts the employees sent by the unions. The construction firm accepts the union's evaluation of the employees' qualifications, unless and until there is evidence to the contrary.

Through the hiring hall the construction firm relies on the union to provide employees in conformance with the various equal employment opportunity statutes, rules, and regulations.

When most of the firm's hourly workers are permanent or near permanent employees, fluctuations in the availability of employees is solved through the use of the hiring hall. There may be a temporary shortage of a crew or of a particular skill. The fastest solution is for the firm to call upon the union. It may not be the cheapest solution (for example, when a union employee puts gasoline in a diesel engine). Regardless of the risks, however, there are no acceptable alternatives.

THE EFFECTS OF SEASONALITY

With the possible exception of parts of Florida and California, the construction industry in the United States is subject to a poor weather season. In the parts of Florida and California that are not subject to winter weather, rain and, in the case of California, the rainy season force construction slowdowns and shutdowns. Most construction firms attempt to plan their projects around the seasonal patterns. The plans include the pouring of concrete and the laying of asphalt in good weather. In addition, construction firms plan the completion of building interiors in the off-season or during bad weather.

The plans of the industry are affected by the demand for and the timing of the start and completion of construction projects. If the contract is awarded at an unproductive time of the year, the plan and projected completion date of the project must be adjusted accordingly. As the productive season comes to a close, there is more and more pressure from customers to complete projects or to complete them to the point that the work can continue in the off-season. The need for accurate planning intensifies.

The effects of the nearing of a project's completion date can be traumatic on the construction firm. The firm's reputation and profit incentives may be contingent on completion within the date specified in the contract. Meeting these dates requires close and accurate scheduling and can cause the hiring of more employees than usual to complete the project. One threat that is always present in the construction industry is union jurisdictional disputes. Because of the number and differences between the unions, jurisdictional questions can dominate the union actions and delay completion of a project. There may be little that the construction firm can do to correct the problem.

Each construction firm must resolve the problem of which employees to retain during the off-season or rainy season. When employees are laid off during adverse weather they may move to other industries or other firms who have work that can continue despite the weather. The firm must identify employees it cannot or does not wish to lose and devise policies and incentives for keeping them. Since these policies are expensive, the firm must rank and rate its employees to identify those that are needed. These employees usually include managers, foremen, and long term and highly productive salaried and hourly employees.

PROJECT DIRECT AND INDIRECT LABOR

The difference between project direct and indirect labor is one of identification. Project direct labor is the labor that can be traced and identified with a single project or phase of a project. This type of labor includes construction crews, equipment operators, and foremen who are assigned to one project or one phase of a project. Indirect labor includes security personnel, timekeepers, construction clerks, and drivers who are assigned to more than one project or more than one phase of a project. The method of recording and classification of labor differs from firm to firm and project to project but the objectives are similar.

RECORDING TIME WORKED

There are two aspects to timekeeping in the construction as well as other industries. One objective is to accurately record the time that employees work. The other is to record where the employees spend the time that they work, that is, the projects or phases on which they spend their work time. It is important that the time worked and the time spent on projects or other functions equal.

In industries other than construction, employees' time worked is recorded on a time clock. Time clocks are unusual at a construction site. The resistance to time clocks at constructions sites is traditional. There is also a practical reason. In an extensive project a loss of productivity would occur regardless of the placement and number of clocks. They would need to be moved frequently along with the time cards as the crews moved from project to project. Normally, time clocks are found in the maintenance shops and at the permanent plant locations such as concrete, asphalt, or washing plants. If the trucks are kept in a central location, there may be a time clock for the drivers.

Without time clocks, field supervision must be relied upon to report time worked by construction employees. Field supervisors report and sign for the time worked by their employees. Therefore, the accurate and timely payment of employees depends on reports prepared and/or approved by their field supervisors.

Manufacturing firms use automated methods for recording the time that employees spend on each job. The matching of the time worked and the time spent is almost automatic, with only manual input for the exceptions. In the construction industry with its widespread geographic locations these systems are almost nonexistent. Project or job cards are rarely used in the construction industry, even with manual systems.

Time recording cards, too, are not generally used on individual construction projects. The industry is not receptive to the use of job cards even if it is necessary to record where the time worked is spent.[1]

---

[1] L. Gayle Rayburn, *Principles of Cost Accounting With Managerial Applications* (Homewood, Illinois: Richard D. Irwin, 1979), pp. 143-171.

JOB OR PROJECT REPORTING

Despite the aversion to the use of job and time cards, it is necessary for the construction firm to know how many hours its employees work and where they spend those hours. The construction firm must know which contract to invoice for the work completed and must, therefore, know to which project the employees were assigned. If the firm does not collect the time worked by project and phase, it will be unable to determine the profitability of the various phases and projects. Historical information will not be available for forecasting and bidding and the firm will be unable to recognize its mistakes and correct them.

The construction firm typically includes reports of where employees worked on the same documents that are used to report the employees' time on the job. This implies the use of one reporting document rather than two. Again, the firm relies on field supervision to approve, that is, verify the projects and phases to which each employee's time is to be charged. From the control and financial point of view, the reporting of the employees' time spent at work and where that time is used depends on field supervison. The field supervisors control the amount of time paid for each employee and the projects and phases of projects against which the time is charged. Field audits can be used to verify the accuracy of this information.

TRUCK HIRE PAYROLLS

A payroll peculiar to the construction industry is the truck hire payroll. The truck hire payroll is prepared for the independent truckers hired by the construction firm. It is prepared as a service by the firm for the independents. Usually, the independent truckers are not charged specifically for this service; however, it is argued that the rates they are paid for hauling have been adjusted to include the costs of administration of their payroll.

MATERIALS TRANSPORTATION

Construction firms generally own their own trucks. However, the number of hauling vehicles owned is not sufficient to provide for the firm's total needs, especially in the peak of the productive season. The degree of use of the firm's own vehicles varies from firm to firm. It is not efficient for the firm to provide 100% of its hauling capacity. That capacity may not be known and any estimates of the amount needed will usually be inaccurate. Therefore, it is accepted practice to hire the additional capacity. "Gypsy truckers" are used to meet the peak needs of the firm.

Two widely used types of truckers are available for hire by the construction firm. One is the hourly hauler—usually a large organization that leases trucks for an hourly rate. The rate includes the driver, the fuel, depreciation, repairs, and a profit for the hauler. The second type of hauler is the independent trucker. The independent trucker owns his truck and is paying for his vehicle loan. The independent trucker pays all of the operating costs, such as gasoline, oil, and vehicle repairs.

The only direct assistance independent truckers receive is the use of the construction firm's buying power. Construction firms will often sell the independent trucker gas, oil, and truck parts at cost or at a discount price. Independent truckers are reimbursed per ton-mile of material hauled. The faster they haul, the more they haul and the more excess cash they earn. When they have earned their fixed costs, excess earnings above variable costs are for their personal use. Compared to the hourly hauler, the independent trucker is more efficient. The independent trucker has sufficient incentive to haul as much material as possible as quickly as possible.

RECORD KEEPING FOR HAULING

For measuring performance, hauling costs must be identified with the appropriate construction project. When the trucker is assigned a load or a series of loads, he (or she) is given a project or phase number to include on the load sheet. This is the basis for the project charge. If, however, the trucker is hauling material at a plant or in the construction yard, a code is included on the hauling sheet that identifies the expenses as part of indirect costs to be allocated to all projects and phases affected.

Because independent truckers are used to the extent that they are available, it is common practice for the construction firm to record the material hauled by the trucker and to prepare a payroll for the trucker, in some cases including payroll deductions. The independent trucker has his load sheet copy receipted upon delivery of the material and submits the receipted copy to the construction firm's financial department (the trucker retains an extra copy). The receipted copy submitted to the financial department is matched with a copy of the original unreceipted load sheet. If the comparison is valid, the gross amount due the trucker is computed by reference to the rate per ton for the distance hauled.

Independent truckers follow the work. In one season they may work for a number of construction firms (thus the name gypsy). Some independent truckers attempt to haul continuously for one construction firm. If the trucker has hauled continuously for one firm, the firm's financial department will not only prepare his form 1099 (summary of earnings of an independent contractor), but will also calculate, withhold, and remit his payroll deductions. For those who haul for many construction firms, the financial department will calculate the gross earnings and prepare the annual form 1099.

ALLOCATION OF OVERHEAD COSTS (INDIRECT COSTS) TO CONSTRUCTION PROJECTS

Overhead costs are those costs that cannot be identified with one specific project or phase of a project. Overhead costs are incurred for services and materials that benefit more than one construction project or phase of a construction project and, therefore, must be allocated to each. These costs include labor, materials, hauling, and equipment.

CLASSIFICATION OF OVERHEAD COSTS

Generally, overhead costs are classified as fixed or variable. Accounting textbooks will add another category described as semifixed or semivariable. Variable overhead costs are the type of costs that change with the number and level of activity of construction projects. Examples are field supervision, construction clerks, loader operators, and possibly crane operators. The classification depends on whether it is necessary to gather cost information by phase as well as by project. Depending on the reporting requirements, the traceability of the costs can be established. For this discussion it is assumed that costs are accumulated by project phases.

Fixed costs do not change with the level of construction activity except in the extremes. Examples of fixed costs are the firm's president's salary, property taxes, insurance on company owned buildings, depreciation on the central office and some items of equipment, and the expenses of the financial, personnel, and legal department. In current practice the designation of these costs as a group is changing from fixed to discretionary.

Discretionary is a more appropriate description because these costs will only change at the discretion of management. Managers must make a decision to change these types of costs. Unlike variable costs, they do not change automatically with changes in construction volume. If a foreman does not need supplies, he will not order them. The financial department may need to be told that it is not needed! To be profitably controlled, the need for discretionary costs must be continuously examined by the construction firm's managers.

Semivariable costs are more difficult to control and classify. With the use of statistical techniques, the fixed and variable components of semivariable costs can be separated on an approximate basis. In practice the information needed may not be available to separate the parts of these costs. In addition, the information available may be inaccurate and the separation may be misleading. These costs may also vary in steps or at only certain levels of activity which cannot be predicted with sufficient accuracy.

The most useful classification of semivariable (semifixed) costs is as discretionary costs. Their importance and effect will then not be ignored because they will be reexamined with other discretionary costs. The continuous reexamination of semivariable costs will support the testing and evaluation of the nature and changes in this cost category. There is less risk to the construction firm from this approach to semivariable costs than from assuming that these costs will change automatically in part or in whole with changes in construction activity.

A complication of the classification of variable costs in the construction industry is that one type of cost may be classified differently for different construction projects. For example, if fuel and trucking costs apply to only one project but to a number of phases of that project and cost by phase is needed for reporting and bidding, these costs will be treated as variable indirect (overhead) costs for one construction project and allocated to all phases of the project (if the costs were only incurred for one phase, they would be classified as direct).

If security, inspection, and construction supplies are purchased and used for a group of construction projects, the costs are accumulated as an additional level of indirect costs and allocated to the various projects. Within the construction projects these costs are allocated to each phase. Costs that are incurred for all construction projects such as security, field general supervision, and leased communication equipment are allocated as indirect overhead to all construction projects. Again, if necessary, within each project these costs are reallocated to each phase of each project.

METHODS FOR ALLOCATING OVERHEAD COSTS

It is not necessary to allocate all overhead costs. An alternative approach is to allocate only those indirect costs that are directly traceable to one or a few construction projects. Deducting these and the direct project costs from the project revenue results in a project contribution that is used to measure the success or failure of the project. The use of the contribution approach is discussed in more detail in the next chapter. For this chapter, however, it is assumed that the decision is to allocate all indirect costs to construction projects and/or phases.

Similar methods can be used for the allocation of indirect variable overhead or for indirect fixed overhead. The same or similar methods apply for the allocation of overhead costs between construction projects or between phases of the same project. But the firms managers must be aware of the methods used to allocate fixed (discretionary) overhead costs because of the effects of allocation on future decisions. That is, regardless of the method used to allocate the president's salary to construction projects, the addition of a project will not affect his (or her) total salary. The most popular allocation methods are construction direct labor hours, construction direct labor dollars, total project costs without overhead and excluding subcontracting, total project costs before the overhead allocation excluding direct project materials but including subcontracting, or total project costs before the allocation of any overhead.

Direct labor hours are a valid method of allocation because construction project costs are usually related to the hours of time worked by construction crews. Crew time affects the need for supplies, for supervision, and for clerical support. The weakness in this method is the need to collect crew hours on a timely and accurate basis. This method is not effective if a major portion of the project is subcontracted. The overhead rate for this method is calculated by dividing the total overhead by the total construction project direct labor hours. The total crew hours for each project are multiplied by this rate to obtain the overhead allocated to that project (or phase).

Another method of overhead allocation is direct labor dollars. This is the most popular method because the information is readily available in the accounting system. The dollars of crew labor are collected for payroll and tax purposes. This method has the same advantages and disadvantages as the use of direct labor hours. The one major difference is that there is no incremental cost for collecting this information, since it is already available. This allocation is expressed as a percentage and is calculated by dividing the total overhead by the total construction direct labor dollars. Each project's direct labor dollars are multiplied by the calculated percentage to obtain the overhead cost for that project.

Allocating overhead based on total project costs excluding subcontracting is another popular allocation method. It is argued that some overhead costs, such as purchasing, receiving, security, and the use of supplies, are related to materials used as well as the number or amount of direct labor. It is also argued that subcontractors provide these services for themselves and therefore should be excluded. The allocation method is calculated by dividing the total construction costs before overhead less subcontracting into the total overhead costs. The resulting percentage is multiplied by the total costs incurred on each project without subcontracting costs, which is the overhead chargeable to each construction project.

The nature of the construction project and contractual provisions influence the selection of the allocation method. If subcontracting costs are a major part of the project costs, a useful allocation method is project total cost before overhead but including subcontracting and excluding project direct materials. This method is particularly useful when most of the work is done by subcontractors who furnish most of the materials. The allocation is calculated by dividing the total firm overhead by the total construction costs excluding direct materials costs. The allocation percentage is multiplied by the total costs of each construction project less any project direct materials costs to determine the overhead allocable to each project. When this method is used, some construction firms will calculate a separate materials overhead for allocating overhead costs related to the purchase, storage, and use of materials. Material overheads are not widely used.

To eliminate any incremental costs and to simplify the allocation method, overhead may be allocated based on the total costs of all projects before the allocation of overhead. Multiplying this percentage by a project's total cost equals the amount of overhead allocable to that project.[2]

The reader must remember that the allocation of overhead costs within a project to phases of a construction project is a common occurrence. The requirement may be contractual or it may be desired by the construction firm. The contractual requirement may be caused by the need to depreciate the different phases of a construc-

[2]*Ibid.*, pp. 213-242.

tion project at different rates for both tax and financial statement preparation. For example, the costs associated with site preparation are part of the cost of the land and are not depreciated, while the costs of land improvements, buildings, and machinery and equipment are depreciable but at differing rates.

The construction firm may wish to accumulate costs by project phases because a future project may be similar to a phase of a past project. Future projects may have phases that were combined differently in past projects. To bid successfully and profitably on contracts and projects in the future, the construction firm must have the historical information from which to prepare bids and estimates.

Overhead applications in a construction firm are more inclusive than in a manufacturing environment. In manufacturing general and administrative costs are not included in the overhead application. In the construction industry they frequently are included. Each construction contract has many unique features. The bids and estimates are based on total costs and success or failure depends on earning an income after all of the firm's costs. To measure performance properly for each project it is necessary to include all applicable costs as either direct or allocated costs.

SUMMARY

Chapter 8 explores the personnel policies of the construction firm. The policies of the construction firm differ from those of the manufacturer. In addition to the usual policies, the construction firm relies heavily on the union hiring halls. This reliance affects the firm's approach to employee continuity and the various equal employment opportunity statutes. The construction firm also has the unusual problem of retention of employees during the off-season or slow season.

Also examined in this chapter is the construction firm's peculiar approach to the recording of employees' time worked and where the time worked is spent. The general lack of use of time clocks and job cards in the construction industry is noted. The result is the almost complete reliance on field personnel to record employees' time worked and its use.

Truck hire payrolls, which are found only in the construction industry, are described. The description includes the two types of hauling services most widely used in the industry and the advantages of each. The discussion emphasizes the use of independent truckers and the services provided by them and to them. The payroll service provided by the construction firm for the independents is stressed because of its importance to the relationship.

The allocation of overhead costs to construction projects is also analyzed. Fixed or discretionary overhead is defined, as is variable overhead. A suggestion for the treatment of semivariable (semi-fixed) costs is included in the analysis. The grouping of costs for allocation to construction projects and phases is discussed.

Allocation methods (bases) are described and the calculation of the allocation factors presented. The description includes the most popular methods, such as project direct labor hours, project direct labor dollars, total project costs, total project costs excluding subcontracting, or total project costs excluding project direct materials.

As a final point, the need for the allocation of costs within projects to phases of projects is mentioned. Included in the discussion are the reasons for cost allocations to project phases, such as contractual obligations or the desire of the construction firm to collect this information.

QUESTIONS

1. Why do the personnel policies of a construction firm differ from those of a manufacturer?
2. How are construction workers hired?
3. Does seasonality effect the construction firm? If so, how?
4. Describe the difference between project direct and indirect labor.
5. Discuss the methods generally used to record employees' time worked and those used in the construction industry.
6. In the construction industry, how are the projects that the employee works on reported?
7. Is it necessary for the construction firm to know where the employee spends his or her time at work?
8. What is a truck hire payroll?
9. Describe the alternatives available to the construction firm for hauling materials?
10. Describe an hourly trucker.
11. Describe an independent trucker.
12. Why does the construction firm prepare a payroll for the independent trucker?
13. Why is it necessary to allocate overhead costs to construction projects?

14. How are overhead costs classified?
15. Define fixed or discretionary costs.
16. Define variable overhead costs.
17. Define semivariable (semifixed) overhead costs.
18. How are variable costs classified for allocation purposes?
19. What overhead cost allocation methods are widely used in the construction industry?
20. Which of the methods used is the most popular and why?
21. Are overhead costs allocated to a construction project, or in more detail?
22. Why are overhead allocations made in more detail than to a construction project?

REFERENCES

Anthony, Robert N., and James S. Reece, *Management Accounting* (Homewood, Illinois: Richard D. Irwin, 1975).

Chorba, George J., *Accounting For Managers* (New York: American Management Association Extension Institute, 1978).

Dearden, John, *Cost Accounting and Financial Control Systems* (Reading, Massachusetts: Addison-Wesley Publishing Co., 1973).

Gordon, Myron J., and Gordon Shillinglaw, *Accounting: A Management Approach* (Homewood, Illinois: Richard D. Irwin, 1974).

Horngren, Charles T., *Cost Accounting: A Managerial Emphasis* (Englewood Cliffs, New Jersey: Prentice-Hall, 1977).

Kieso, Donald E., and Jerry J. Weygandt, *Intermediate Accounting* (New York: John Wiley & Sons, 1977).

Kieso, Donald E., and Jerry J. Weygandt, *Intermediate Accounting* (New York: John Wiley & Sons, 1980).

Meigs, Walter B., A. N. Mosich, and E. John Larson, *Modern Advanced Accounting* (New York: McGraw-Hill Book Company, 1979).

Montgomery, A. Thompson, *Managerial Accounting Information* (Menlo Park, California: Addison-Wesley Publishing Company, 1979).

"Professional Notes," *The Journal of Accountancy*, December 1979.

Pyle, William W., John Arch White, and Kermit D. Larson, *Fundamental Accounting Principles* (Homewood, Illinois: Richard D. Irwin, 1978).

Rayburn, L. Gayle, *Principles of Cost Accounting with Managerial* Applications (Homewood, Illinois: Richard D. Irwin, 1979).

Welsch, Glenn A., Charles T. Zlatkovich, and Walter T. Harrison, Jr., *Intermediate Accounting* (Homewood, Illinois: Richard D. Irwin, 1979).

Weston, J. Fred, and Eugene F. Brigham, *Managerial Finance* (Hinsdale, Illinois: The Dryden Press, 1978).

CHAPTER 9

CONSTRUCTION PROJECT (JOB) COSTS AND BUDGETING

An important contributor to the success of the construction firm is the collecting of costs by construction project and the plan for the expectations from each project. Although it is very rare for two projects to be exactly alike, historical information is the beginning of planning. Historical information is adjusted for the uniqueness of the future project.

COLLECTING CONSTRUCTION PROJECT COSTS

The assigning of project numbers is the vehicle for collecting project costs. The project number is used for collecting direct material, project direct labor, subcontracting, and all indirect costs. If costs are collected by project phase rather than total project, within the project number a section of digits are reserved to identify phases.

ASSIGNING PROJECT NUMBERS

Project numbers are assigned in sequence by the fiscal year in which the project is initiated. The number assigned should be long enough to include all the projects that are initiated in one fiscal year. The project number is assigned sequentially within each fiscal year and should have enough digits to identify phases of construction projects. For example, the project number may have eight digits. The first two digits identify the year in which the project was initiated such as 80 for 1980. Assuming that the largest number of projects started in one year does not exceed 999, the next three digits are used for the sequential project number, with the first project assigned the

134

number 001, the second project initiated in 1980 the number 002, and so on. The last three digits of the project number are reserved for phases. The numbers for the first project in fiscal year 1980 with two phases would be 80001001 and 80001002.

All cost information is reported by the project number. Costs are collected and reported for the current period, for the fiscal year to date, and as cumulative costs since the project was started. The assignment of project numbers is controlled by the financial department. The financial department cross-references the project number to each contract and verifies that all contracts (projects) are assigned a number. Control is important because of the effect of project costs on profits, cash flow, and planning for the future.

Project numbers may be subdivided for more than the collection of costs per phase. Additional digits may be included to collect and summarize cost categories, such as direct labor, direct material, subcontracting, and indirect or allocated costs. A one or two digit indicator can be used to identify these costs for each project. For example, using the project number from above, assume that the project direct labor is coded 01, project direct material 02, project subcontracting 03, and indirect costs 04. The complete project number would be 80001002-02 for direct material, 80001002-03 for project subcontracting, and so on. Any desired information can be collected about all or any one project in this manner. A code could be assigned for equipment usage, supplies, and truck hire.

RECORDING PROJECT COSTS

Once the project number has been assigned, it must be used throughout the accounting system if the information reported is to be of any use for recording and analyzing project costs. To properly record the costs for project direct material, every purchase order for project material, that is, material that is used directly for one or a few projects, must have the project number included on the purchase order. If the materials is for stock or for indirect use, the purchase order must have the inventory account number or the indirect material account number on it. A purchase order must not be issued without one of these numbers included in it.

If the material is purchased for inventory for later use, the document that removes the material from inventory must include either a project number or an indirect account number. A material requisition should not be filled, without these numbers.

Subcontracting costs must be treated like material costs. Each subcontract invoice, including accrued invoices, must contain the project number for which the work was performed. The agreement with the subcontractor must require the inclusion of a project number on each invoice to the construction firm. Because subcontracting costs are usually large, it may also be necessary for the subcontractor to include a phase identification. When the subcontractor works on a number of projects, the construction firm personnel may need to allocate the invoices among the various projects on an estimated basis.

For project direct labor field personnel must include the project or account number to which the hours apply on the time card in the "where worked" section. Because employees can be sick, on vacation, absent with permission, or used for indirect work, the job card or job section must contain an account designating one of these conditions, or a project number. Time cards that do not account for the employees' time should be returned to the field personnel for completion. To reduce the costs of administration and the effort required by field personnel to complete the project information, time should not be allocated for less than one hour. For example, an employee should report time spent that is less than an hour together with the time worked for the remainder of the day.

Indirect costs must be identified by an account or numerical designation. There can be a number of indirect cost identifications, depending on whether the cost is allocated between phases of one project or between two or more projects. Costs that are classified as indirect vary from project to project and firm to firm.

CLASSIFYING CONSTRUCTION PROJECT COSTS

The exact classification of construction project cost varies from firm to firm. However, certain criteria are used to classify costs as direct or indirect and within the indirect category. To the extent possible, the firm should classify costs as direct and charge them directly to the project.

DIRECT COSTS

Construction project direct costs are such costs as project materials, project labor, subcontract costs, equipment charges, and truck hire. Project direct materials are materials that are used in the construction project and are part of the final product be it a building, a road, a foundation, or a house. Materials that are usually classified as direct are concrete, steel and wood support beams, wallboard, and windows.

Other materials such as nails or fasteners are part of the finished product but their costs may be less than the costs of assigning the amounts used to each product. As in all accounting decisions the guide is the principle of materiality. Only direct materials that are material in amount should be charged directly to a construction project.[1]

Project direct labor is labor cost that is used solely on one project and can be identified as such. The labor is charged to the project on the job section of the time card by project number. Employees who work directly on the project are classified as direct labor. Examples of direct employees are concrete crews, framing crews, road crews, and finishing crews, as well as their immediate supervision. Again, materiality must guide the degree of detail into which employees' time is divided.[2]

Subcontract costs are usually incurred for one or a few projects and the work for each project can be accurately defined. The subcontractor must include the project number and description on the invoice to the construction firm. In some cases, the work may be applicable to more than one project. If the subcontractor cannot identify the portion applicable to each project, employees of the construction firm must allocate the charges to the projects involved.

Often equipment usage can be traced to a construction project and can be classified as a direct cost. Field personnel should report the hours of equipment usage for each project. When the proper rate is applied, this becomes the direct charge for equipment usage for that project. If the hours of use cannot be traced to each project, equipment costs must be allocated as part of indirect costs. The accounting principle of materiality is applicable to this decision.

As stated above, truck hire can be and usually is incurred for the completion of a project. Hauling is usually related to the grading and site preparation including parking lots and streets. Where possible, truck hire should be identified with the project and included as a direct cost. Charging costs directly is important for reimbursement, profitability, and planning for future projects. Including the project numbers on the hauling sheets will correlate the project and the hauling charges. Hauling of a general nature, such as within the construction plant or yard, and hauling that will not materially effect project costs can be charged to overhead and allocated to all construction projects.

---

[1] L. Gayle Rayburn, *Principles of Cost Accounting with Managerial Applications* (Homewood, Illinois: Richard D. Irwin, 1979), pp. 29-30.
[2] *Ibid.*, pp. 29-30, 219-220.

INDIRECT COSTS

Overhead or indirect costs are costs that are allocated to construction projects. Such costs are supplies, construction clerks, security, and other expenses that cannot be traced directly to one project or one phase of a project. If the costs incurred are to be allocated to the phases of a construction project but cannot be traced to each phase, they will be charged to the project and allocated to the project phases on a basis such as those suggested in the preceding chapter. Costs of this nature could include drivers, security, clerks, and supervision.

There are also indirect costs that apply to more than one construction project. Costs of this nature include, but are not limited to, general construction clerks, drivers, field supervisors, field superintendents, insurance, and store clerks. These costs are allocated to construction projects and then to phases of construction projects, if necessary.

Field overhead that relates to all construction projects includes the operations supervisor and staff, the purchasing agent, and expeditors. Field overhead is allocated to all construction projects and, within projects, to project phases.

Home office or central overhead costs are allocated to all construction projects. This approach to central overhead differs from the manufacturing approach. Central overhead may be allocated after the fact in a manufacturing environment to measure total profitability; however, prices are normally market determined. But the construction firm computes the price as a factor above the total project costs. Therefore, the total costs of the firm must be allocated to all of the construction projects or losses will be incurred. Central overhead costs include the president of the firm and his or her staff, accounting and finance, personnel, legal, and estimating and bidding.

REPORTING CONSTRUCTION PROJECT COSTS

Reporting on the costs of construction serves more than historical purposes. It allows management to adjust the costs and performance on the project, if necessary, from poor or mediocre to acceptable. Therefore, reporting must be timely. If the reports are not timely, the accountant becomes solely a historian. The managers of the firm can only cry over poor performance reported too late for them to act.

PERIODIC REPORTING

Reporting intervals must be as short as possible. The shorter the reporting interval the higher the probability that the firm's managers can react to unfavorable information. Weekly reports should be prepared for managers who are in a position to react to the reports. Often weekly information is not available in complete form. Therefore, weekly reports should be prepared with approximate or estimated information. For example, the payroll hours by labor category are multiplied by estimated or average labor rates to approximate the direct labor cost. The direct labor cost is multiplied by an estimated overhead rate to calculate the overhead cost. The direct material cost can be estimated from the material used and the subcontractor charges can be estimated from the work done that week. As long as these estimates do not materially differ from the actual costs, they are sufficient for management to act upon.

Weekly information must be summarized into monthly project cost reports. Monthly information is usually less volatile than weekly information and may, therefore, be more useful to management. But the report must be issued within a very brief interval after the end of the month. The publication of a May report in July, for example, will be of little value to management because the information is stale. If all the information required is not available on a timely basis, estimates should be used to permit the publication of the monthly report.

Quarterly reports may also be useful because they reduce random variations. The quarterly reports should be less subject to random interruptions than the monthly reports and therefore more useful than the weekly or monthly reports. Again, the quarterly reports should be issued as soon after the end of the quarter as possible.

Project cost reports should also include a total for the fiscal period by project. Although this information is less timely and therefore less useful, the firm's managers may be able to change the course of events based on the information received. Because of this possibility, the fiscal year cumulative report should be issued as soon after the end of the reporting period as possible.

ACCUMULATED REPORTS

In addition to weekly, monthly, and quarterly reports, an effective reporting system should include accumulated totals. Cumulative totals normally reduce the effect of random variations in the data reported. Accumulated reports should be prepared as soon after the end of the reporting period as possible. The cumulative report should be prepared for each interval report and preferably should include a comparison with the objective.

Accumulated project cost reports are usually prepared for each week of the fiscal year, for each month of the fiscal year, for each quarter of the fiscal year, and, as noted above, for the fiscal year to date. Accumulated reports differ in the degree of estimates employed and the presence of any unusual or extraordinary items.

Some construction projects require more than one fiscal year for completion. An accumulation should then be included for the project to date, that is, since the start of the project. This report should also include a comparison to the objective for that project. In these reports, as in all others, estimated rates and applied overhead must be used if needed to issue the reports on time. An example of interval and accumulated reporting is included as Exhibit 11. The use and reporting of objectives are illustrated and discussed later in this chapter.

TYPES OF CONSTRUCTION PROJECT COST REPORTS

In the preceding discussion of interval reporting, it was assumed that cost reports were prepared by construction project. In addition, it was assumed that the reports were prepared with the full allocation of all costs. However, the types of reporting can vary depending on the style and the needs of the management of each construction firm. For example, reports may be prepared by each phase of each construction project, by contract (which may include more than one project), or by customer.

Reports can also be prepared before or after income taxes and with only variable project costs, emphasizing contribution margin rather than net income. The reports may also include comparative data such as a previous period, a previous project, a budget, a contract amount, or an estimate or bid.

REPORTING BY PHASE, CONTRACT, OR CUSTOMER

Circumstances, as noted above, can require the construction firm to report by phase. For this type of report costs are accumulated by cost element for each phase of a project that has either been defined in the contract or by the managers of the construction firm. To ensure accuracy, phase reports should equal the report for the total construction project or projects. These reports may be prepared for any time interval desired.

Construction contracts often require that more than one project be assigned because of the nature of the construction specified by the contract. As a result, project reports may need to be combined. Such reports are usually prepared at various time intervals and are always checked with the individual project reports to verify the accuracy of the summary report.

A construction firm may have a number of contracts with one

customer. Consequently, the construction firm will have a number of projects with the same customer. To evaluate the customer's performance management would need a summary report reflecting the total performance on this customer's projects. If such a report is prepared, it should be prepared as soon as possible after the end of the reporting period and should be checked against the individual project cost reports to verify the accuracy of the report. An example of this type of cost report is included as Exhibit 12.

COST REPORT CONTENT

Whether to report project net income before or after income taxes on the cost or performance report is a choice for the firm's managers. Some managers feel that they do not directly control income tax rates and therefore are not responsible for the tax amount. Other managers want the reports to reflect the final contribution to the firm of the project and therefore require the reporting of net income after taxes for each project. The most common type of report reflects net income before taxes.

To report the net income after tax by project, the project net income before taxes is adjusted for the firm's estimated average tax rate, which is adjusted for any effect from that project such as the tax effect of an extraordinary gain or loss. Whichever method is used, the reports must be clearly labeled to prevent misinterpretation.

Project performance or cost reports that are made up to report net income before taxes can be prepared on a fully absorbed or a contribution basis. Reports prepared on a fully absorbed basis include the total cost of the construction firm whether direct or allocated. Unlike manufacturing, general and administrative costs (also known as period costs) are included in the allocation. The reported net income before income taxes is net of all costs for the firm and reflect the project (or phase, etc.) net income to the extent of the accuracy of the cost allocations.

Cost reports prepared on a contribution basis do not include the firm's total costs identified by project. The revenue for each project (or phase, etc.) is listed. Project direct materials and direct labor are deducted from the revenue to calculate net revenue after prime costs. The variable overhead costs are listed next. Variable overhead from this project alone is added to variable overhead shared by two or more projects. This total is deducted from the net revenue after prime costs to calculate the project contribution.

The project contributions are summed, and from this sum the firm's fixed or discretionary costs are deducted to identify its operating net income before taxes. With this method the fixed and semifixed costs (discretionary costs) are not allocated to construction projects, but are deducted in total to compute net income before income taxes. This approach is particularly useful and applicable to construction firms whose projects are competitive in nature and of standard or near standard construction, such as tract housing. This approach is also applicable during periods of slow construction activity when a construction firm may accept a contract with only a contribution income, not expecting to recover full costs and earn the usual amount of net income before income taxes.[3] An illustration of a project contribution report is included as Exhibit 13.

Construction project cost (performance) reports must include comparative data to enhance their usefulness. The types of data available are budgets, bids or estimates, contract amounts, and prior periods information. The most useful comparative item is the budget or objective for the project. However, a project budget may not be available. The bid or estimate from which the bid was prepared may be the only and the best source of comparative information, especially if the bid or estimate has been updated for contract amendments.

The contract itself may be a source of comparative information. If the detail cost information is part of the contract, the firm can use it as the project objective. However, the data must be updated for contract amendments and changes. In the construction industry, historical information is less useful than other types of data. Projects are rarely exactly alike. Unless the projects are of standard construction or information can be gathered from the phases of different projects, historical data are an approximation at best and must be used cautiously.

Comparative data should be included for each interval reported and for each interval accumulation. If the week, month, and quarter cumulative data are included, a cumulative comparison should be reported through each period. In addition to the financial information, a brief description of the project or work must be given. The descriptive information influences the interpretation of the financial data. The work description tells the reader whether the project's materials are scarce or difficult to use, for example. The description also discloses any peculiar construction problems inherent in the job. Exhibit 14 is an example of a comparative report.

CONSTRUCTION BUDGETING

Budgets and planning are important to the success of the construction firm. However, budgeting is extremely difficult because of the

---

[3]*Ibid.*, pp. 506-507.

effect of economic fluctuations and the firm's dependence, in large segments of the construction industry, on requests for proposals which may be received without prior notice. One solution is to shorten the budgeting period from a year to a few months. But even a short budget must be evaluated in terms of probabilities and may be inaccurate.

The types of budgets needed by the construction firm are revenue and expense budgets, capital asset budgets, and cash flow budgets. If possible, the revenue and expense budgets should be detailed by project (and in some cases phases), by contract, and by customer.

REVENUE AND EXPENSE BUDGETS

Project budgets are extremely important and necessary for controlling costs and supporting profits. Because the responsibility for performance is normally assigned to field personnel on a project by project basis, measuring the performance of field personnel through each project is vital. Project direct and indirect material can be estimated by walking through the bid and adding a contingency for the unexpected. This is necessary to order the material needed for the physical completion of the project; therefore, pricing the material is the only additional step for a project materials budget.

If part of the project is to be subcontracted, the construction firm solicits bids or estimates from its subcontractors. The bids selected or the estimates can be used to calculate the subcontracting budget for the project. Within some circumstances, subcontract proposals or bids are used as the basis for estimating project work to be done by the construction firm and, therefore, can be used as budgets.

A walk through the bid can identify the hours of labor by type as well as the hours of estimated equipment usage and the amount of hauling needed for the project. When rates are affixed to these estimates, the result is a budget. The estimates prepared from the bid include materials and indirect labor. These costs are part of the overhead to be assigned to each project that enters into the total project cost.

The financial department must project the remainder of the overhead costs for the budget period. When these costs are added to estimates from the bid analysis, the total overhead budget for the period is available. The next step is to allocate the budgeted overhead to the construction projects. The same allocation base is used for allocating the budgeted overhead as for the actual overhead.

Some firms use the budgeted overhead and allocation method as the costs to be assigned to each project. Each quarter or at least at the end of the fiscal year, the estimated overhead that has been "applied" to construction projects is compared to the overhead actually incurred. The difference, if any, is either allocated to inventories and cost of sales or, if not material in amount, written off to cost of sales. Although the use of applied overhead is more common in manufacturing than in construction, this method of costing overhead to projects is used by some large construction firms.[4]

When the budget or objective has been prepared for each project, it is the basis for other categories of budget reporting. Contract budgets can be calculated from project budgets by adding the budgets for projects that are covered by one contract. If more than one contract covers a single project, the project budget can be disaggregated into phase budgets by estimating the percentage of work that each phase is of the total project. If a budget by customer is sought, the project budgets applicable to that customer can be totaled. Exhibit 15 is an illustration of a project revenue and expense budget.

CASH FLOW BUDGETS

The construction industry's continuous need for cash was noted above. Because of this need a cash budget is essential. Although the flow of cash is difficult to predict because of payment delays by owners, general contractors, and governments, the construction firm requires at least an estimate of the inflow and outflow of cash. If the timing of the inflows and outflows is not compatible, the firm must be able to arrange financing to meet its obligations.

In the past when interest rates were 4 to 6%, banks were willing to retain funds to supply the unexpected needs of clients. Today, it is different. Banks want to know in advance when the funds are going to be required to eliminate any idle reserves. And if the firm cannot predict, the bank may charge interest for funds reserved (line of credit), even though not used by the firm, to compensate the bank for being forced to make a short term investment at a lower rate than a long term commitment.

The construction firm that is unable to predict its cash needs faces increased costs of borrowing, if any funds are available at all! The firm must decide whether the investment in a cash forecasting system is less than the additional borrowing costs without a system.

The first step in the preparation of a cash forecast is the cash

[4]Charles T. Horngren, *Cost Accounting: A Managerial Emphasis* (Englewood Cliffs, New Jersey: Prentice-Hall, 1977), pp. 136-138, 126-136.

inflow by month (a shorter period such as a week is preferable if the necessary information for preparing the forecast is available). Cash inflow is predicted from the estimated collection of accounts receivable. The accounts receivable should be aged as explained in Chapter 6. Historical collection percentages are then applied to the categories of receivables by age, such as government receivables that are over 60 days past due. By applying the percentages to these amounts the expected receipts for each month can be calculated. However, with the changing interest rates and economic activity of the 1970s, it is more effective to express the potential receipts in ranges rather than in discrete amounts and apply probabilities to each amount within the range or at least to the high, low, and median amount in each range, selecting for the forecast the amount with the highest probability of collection. Any expected proceeds from the sale of stocks, bonds, equipment or other assets, and bank loans should be added to collections to determine total cash inflow by month.

The revenue and expense budget, adjusted for accruals and other noncash items, represents the amount of cash outflow for each month. Added to this amount should be expected divident payments, bond retirements, bank loan repayments, and equipment purchases. Deducting the cash outflow from the cash inflow results in the estimated cash balance for each month. Should this balance be negative, financing must be arranged or cash inflow accelerated or outflow delayed.[5] An example of a cash forecast (budget) is included as Exhibit 16.

CAPITAL ASSET BUDGETS

In addition to the revenue and expense and cash flow budget, the construction firm must prepare a capital asset budget. For the firm to complete its projects, the types of assets required must be available when needed. The capital asset budget predicts the need and provides the resources for acquisition of the equipment.

The preparation of the capital asset budget begins with a survey of manager's needs. The survey indicates when as well as what equipment is needed. The survey results are totaled and the total evaluated. If the amount is too high, the budget is reviewed for priorities and adjusted accordingly. Because the costs of the assets included in the budget are estimated, a contingency of 5 to 10% is usually added. The budget is then presented to the board of directors

[5]*Ibid.*, pp. 130-131.

for approval. After approval the capital asset budget is integrated with the revenue and expense and cash budget.

The capital asset budget includes the description of the asset, justification for purchase, the estimated purchase price, and the identification of the user group. The approval process incorporates an estimate of the availability of acquisition funds for each asset. The approval process does not include an evaluation of options of leasing or buying equipment. That decision is reached at a later date. Approval acknowledges need and the availability of financing for either a lease or purchase of the equipment or asset.[6] A capital asset budget is illustrated in Exhibit 17.

SUMMARY

This chapter concentrates on construction project (job) cost (performance) reports and project budgeting. Included is a review of the assignment of project numbers and the collection of project cost information. The review identifies methods for collecting cost information from construction system documents.

The classification of project costs for inclusion in performance reports is examined. Project direct material is defined, as is project direct labor. Also defined are subcontracting and indirect (overhead) costs. The assignment levels and allocation of types of indirect costs are discussed, such as overhead applicable to project phases, to only one project, or to two or more projects and control or "home office" overhead costs.

Alternative reporting periods are explored in this chapter. Preparation of weekly, monthly, quarterly, and accumulated reports is emphasized. Also emphasized is the need for timely reporting, utilizing estimates, if necessary, for each reporting interval, if the results are sufficient for use in making decisions.

Various types of construction reports can be prepared. The types of reports suggested include performance reports by phase, project, contract, and customer. In addition, performance reports may be prepared with net income after income taxes or with project results before income taxes. Reports can be prepared after the deduction of full costs or with only variable costs deducted from project or customer revenue. Performance reports can include comparison to objectives such as budgets or prior periods results.

To be used to evaluate project performance, the necessary budgets must be defined. The definitions of the revenue and expense budget, the cash flow budget (forecast), and the capital asset budget are included in this chapter.

[6]*Ibid.*, pp. 153, 234.

EXHIBIT 11. Project #801001 Performance Report Through July 31, 1980

Cost Element	Latest Week	Week for Month to Date	Latest Month	Month Cumulative for Quarter	Last Quarter	Fiscal Year to Date
			(In thousands)			
Direct materials	$35	$120	$121	$121	$300	$480
Direct labor	10	35	38	38	120	300
Subcontracting	5	20	30	30	90	140
Asset usage	0	4	6	6	12	30
Hauling	1	5	5	5	15	35
Overhead:						
This project only allocated at 100$ of D/L	2	3	3	3	9	21
Central—50% of D/L	10	35	38	38	120	300
D/L	5	18	19	19	60	150
Total project costs:	$68	$240	$260	$260	$726	$1456

NOTES:

(1) This exhibit assumes that the construction firm's fiscal year and the calendar year are the same.

(2) To expedite issuance of this report, the costs reported for the latest week, the week for the month to date, and the monthly cumulative for the latest quarter include estimates. The costs reported for the last or previous quarter and the fiscal year to date are actual (classifying the use of an applied overhead rate as actual cost).

(3) If the project reported extended beyond one fiscal period, the performance report could include an additional column for project costs to date.

(4) The project description could be added to the report heading, such as "paving ten miles of interstate highway in northern Virginia with asphalt" or "construction of a five story office building and accessories (parking lots, etc.) on sand and hard rock at Arlington, Virginia."

(5) Project net income before income taxes can be calculated by adding project revenue to the report and deducting total project costs from the project revenue.

EXHIBIT 12. Customer Performance Report Through July 31, 19XX

Projects	Latest Week	Latest Month	Latest Quarter	Fiscal Year To Date
		(In thousands)		
Project #801001—2 floor commercial building	$ 68	$260	$726	$1,456
Project #791034—ground level warehouse	11	23	44	213
Project #801016—3 level parking garage	58	390	687	687
Project #801003—11 floor office building	98	392	1,123	4,642
Customer total	$235	$1,065	$2,580	$6,998

NOTES:

(1) In this exhibit it is assumed that the construction firm's fiscal year and the calendar year are the same.

(2) To expedite the issuance of this report, the costs reported for the latest week and for the latest month include estimates. The costs reported for the latest quarter and for the fiscal year to date are actual (including applied overhead).

(3) For projects that were in progress for more than one fiscal year, an additional column can be added to the report to include the project total cost to date.

(4) This report is a summary of the total costs of all projects for a particular customer.

(5) Net income before income taxes for each customer is calculated by adding the project revenue for each customer and deducting total cost.

EXHIBIT 13. Project Performance Report Prepared On A Contribution Basis For The FY To Date Through July 31, 1980

Cost Element and Revenue	Project #801001	Project #801002	Project #801003	Total
	(In thousands)			
Project revenue	$3,560	$4,500	$7,040	$15,100
Project costs				
Direct materials	480	300	500	1,280
Direct labor	300	500	700	1,500
Subcontracting	140	200	400	740
Asset usage	30	100	300	430
Hauling	35	100	400	535
Direct overhead: project	21	150	500	671
Total variable costs	$1,006	$1,350	$2,800	$ 5,156
Project contribution income	$2,554	$3,150	$4,240	$ 9,944
Indirect overhead:				
Allocated overhead				1,100
Central overhead				2,500
Total				$ 3,600
Net income before income taxes				$ 6,344

NOTES:

(1) A contribution performance report can be prepared for customer, phase, or any other grouping desired.

(2) This report is prepared for the fiscal year to date; however, a contribution report can be prepared for a week, a month, or any other time period.

(3) For this report it is assumed that the construction firm has only three projects.

(4) Total project revenue has been allocated for the reporting period.

(5) The total column of this report can be extended to include income taxes and net income after income taxes.

EXHIBIT 14.   Comparative Performance Report Through July 31, 1980

Project Number	Latest Week	Latest Week Budget	Variance from Budget (Unfavorable)	Latest Month	Latest Month Budget	Variance from Budget (Unfavorable)	Actual Fiscal Year to Date	Budget Fiscal Year to Date	Variance from Budget (Unfavorable)
				(In thousands)					
801001	$ 68	$ 60	$ (8)	$ 240	$ 300	$ 60	$1,456	$1,500	$ 44
801003	98	100	2	392	390	(2)	4,642	4,800	158
801016	58	55	(3)	390	500	110	687	710	23
791034	11	10	(1)	23	20	(3)	213	200	(13)
Total cost	$235	$225	$(10)	$1,045	$1,210	$ 165	$6,998	$7,210	$ 212
Project revenue	$300	$350	$(50)	$1,200	$1,400	$(200)	$7,300	$8,000	$(700)
Net income before income taxes	$ 65	$125	$(60)	$ 155	$ 190	$ (35)	$ 302	$ 790	$(488)

NOTES:

(1) This report may be prepared by customer, phase, or any other grouping that is needed.

(2) It is assumed that the fiscal year and the calendar year are the same for this construction firm.

(3) The time intervals included in the report are for illustration only; any time interval necessary may be included.

(4) For projects that exceed one fiscal year, columns can be added for the actual and budgeted costs to date.

(5) The project revenue for each interval is calculated from either engineering measurements of completion or estimates of the percentage of the work completed for each project.

(6) Although the costs reported, with one exception, for each time interval are less than budgeted (favorable variation), the project revenue for each interval is unfavorable (less than expected). There was a favorable total project cost variation because the activity levels expected were not achieved. The result was an unfavorable effect on project net income before income taxes.

EXHIBIT 15. Project Revenue And Expense Budget for the First Six Months of Fiscal Year 19XX

Description	Month 1	Month 2	Month 3	Month 4	Month 5	Month 6	Total First Six Months
				(In thousands)			
Revenue							
Projects	$100	$135	$150	$170	$250	$300	$1,105
Other (miscellaneous)	20	20	20	20	20	20	120
Total	$120	$155	$170	$190	$270	$320	$1,225
Expenses (costs)							
Direct materials	$ 10	$ 20	$ 20	$ 20	$ 20	$ 20	$ 110
Direct labor (project)	50	70	80	90	150	220	660
Subcontracting	10	10	10	10	10	10	60
Asset (equipment) usage	20	20	20	20	20	20	120
Hauling	10	10	10	10	10	10	60
Direct project overhead							
Security	5	5	5	5	5	5	30
Project clerks	2	2	2	2	2	2	12
Project supervision	4	4	4	4	4	4	24
Structure depreciation	3	3	3	3	3	3	18
Miscellaneous direct O/H	$ 1	$ 1	$ 1	$ 1	$ 1	$ 1	$ 6

EXHIBIT 15. (Continued)

Description	Month 1	Month 2	Month 3	Month 4	Month 5	Month 6	Total First Six Months
				(In thousands)			
Total project direct costs	$115	$145	$155	$165	$225	$295	$1,100
Allocated overhead at 10% of direct labor	5	7	8	9	15	22	66
Central "home office" overhead allocated at 5% of direct labor	2	3	4	4	8	11	32
Total project cost	$122	$155	$167	$178	$248	$328	$1,198
Net income (loss) Before income tax	$ (2)	$ -0-	$ 3	$ 12	$ 22	$ (8)	$ 27

NOTES:

(1) The revenue and expense budget can be prepared for any time intervals desired, that is, quarters or years.

(2) For this exhibit it is assumed that the firm's fiscal year and the calendar year are the same.

(3) The projected revenue is prepared from estimates of the percentage of project completion for each period or from engineering estimates of measured completion from milestone or other control charts.

(4) The final or accepted revenue and expense budget is the result of a number of iterations.

EXHIBIT 16. Cash Flow Forecast For the First Six Month of Fiscal Year 19XX

Description	Month 1	Month 2	Month 3	Month 4	Month 5	Month 6	Total First Six Months
			(In thousands)				
Cash inflow							
From trade accounts receivable	$123	$120	$137	$140	$150	$120	$790
From loans	-0-	-0-	-0-	-0-	-0-	100	100
From miscellaneous receivables	10	10	10	10	10	10	60
Total inflow	$133	$130	$147	$150	$160	$230	$950
Cash Outflow							
Expenses from revenue and expense budget adjusted for noncash items such as depreciation, amortization, accruals, bad debts, inventory issues, and losses	$ 60	$100	$ 90	$110	$ 80	$130	$570
Repayment of bank loan including interest	100						100

153

EXHIBIT 16. (Continued)

Description	Month 1	Month 2	Month 3	Month 4	Month 5	Month 6	Total First Six Months
			(In thousands)				
Purchase of equipment		40					40
Purchases of inventory			20		20	20	60
Miscellaneous	5	5	5	5	5	5	30
Total outflow	$165	$145	$115	$115	$105	$155	$800
Net cash flow (unfavorable)	(32)	(15)	32	35	55	75	150
Beginning Cash Balance	$ 50	$ 18	$ 3	$ 35	$ 70	$125	$ 50
Ending Cash Balance	18	3	35	70	125	200	200

NOTES:

(1) The cash flow forecast can be prepared for one year or longer if desired. The forecast can also be prepared for weekly rather than monthly time intervals.

(2) The cash inflow from trade receivables is based on the aging of the projected receivables balances at the end of each month.

(3) If desired, an additional outflow for unexpected contingencies could be included in the schedule.

(4) At the end of any month, had the ending cash balance been negative, the firm would have been required to arrange for a loan to cover the lack of cash on hand.

EXHIBIT 17. Capital Asset Budget For the First Six Months of Fiscal Year 19XX

Description	Month 1	Month 2	Month 3	Month 4	Month 5	Month 6	Total First Six Months
				(In thousands)			
Purchases of							
Office equipment			$120				$ 120
Shop equipment				$ 50			50
Shop tools	$ 5	$ 5	$ 5	$ 5	$ 5	$ 5	$ 30
Manufacturing machinery		1,500					1,500
Vehicles					10		10
Building construction	100	100	100	100	100	100	600
Subtotal	$105	$1,605	$225	$155	$115	$105	$2,310
Contingency—10%	11	161	23	16	12	11	234
Total	$116	$1,766	$248	$171	$127	$116	$2,544

Project analysis

Project #	Description	Purchase Date	Amount
1980-001	Office desks, chairs and calculators	March 19XX	$ 120
	Contingency—10%		12
	Total		$ 132

155

EXHIBIT 17. (Continued)

Project analysis

Project #	Description	Purchase Date	Amount
1980-002	Two Pratt & Whitney Machine Tool Milling Machines @ $25	April 19XX	$ 50
	Contingency—10%		5
	Total		$ 55
1980-003	Miscellaneous shop tools purchased equally throughout the period	All Months	$ 30
	Contingency—10%		3
	Total		$ 33
1980-004	One P. & W. Stamping Machine, including installation, shipping, training, and foundation	February 19XX	$ 500
	One portable concrete plant	February 19XX	500
	Two Euclid front end loaders	February 19XX	500
	Contingency—10%		150
	Total		$1,650
1980-005	Three supervisory sedans with radios	May 19XX	$ 10
	Contingency—10%		1
	Total		$ 11

1979-089   Construction of new headquarters building on lot #46 at Main and 8th
            streets

                                                    Construction
                                                    continues
                                                    throughout
                                                    all months. $ 600
                                                                         60

            Contingency—10% $ 660
            Total

Grand Total $2,544

NOTES:

(1) There may be a slight difference in the totals between the two schedules because of rounding.
(2) Each project must also be supported by a detailed description of the calculation of the estimated costs and the reasons for the project (expected benefits).
(3) Based on the capital asset budget, the required payments must be included in the cash flow budget.
(4) The depreciation expense for new assets purchased is included in the revenue and expense budget based on the capital asset budget.

QUESTIONS

1. What is the code required for the control of project costs?
2. Who should control the issue of project numbers?
3. How should project numbers be subdivided?
4. On what documents should project numbers be recorded?
5. How should construction project costs be classified?
6. Define project direct material.
7. Define project direct labor.
8. Define project overhead.
9. What is central or home office overhead?
10. How often should construction project cost (performance) reports be issued?
11. What totals should be included in construction project performance (cost) reports?
12. How many types of construction project cost (performance) reports can be issued?
13. Should construction project cost (performance) reports be summarized before or after income taxes or both?
14. Describe a construction project performance (cost) report prepared on a "fully absorbed" basis.
15. Describe a construction project performance (cost) report prepared on a contribution basis.
16. What comparative information should be included in a construction project cost (performance) report?
17. Should the performance report include work descriptions?
18. What types of budgets are needed for construction budgeting?
19. Describe a construction revenue and expense budget.
20. Describe a construction cash flow budget.
21. Describe a construction capital asset budget.
22. How should a revenue and expense budget be subdivided?
23. How are the revenue and expense budget cost elements estimated?
24. What is the relationship between these three different budgets?

REFERENCES

Anthony, Robert N., and James S. Reece, *Management Accounting* (Homewood, Illinois: Richard D. Irwin, 1975).

Chorba, George J., *Accounting For Managers* (New York: American Management Association Extension Institute, 1978).

Coombs, William E., and William J. Palmer, *Construction Accounting and Financial Management* (New York: McGraw-Hill Book Company, 1977).

Dearden, John, *Cost Accounting and Financial Control Systems* (Reading, Massachusetts: Addison-Wesley Publishing Co., 1973).

Garrison, Ray H., *Managerial Accounting, Concepts For Planning, Control, Decision Making* (Dallas, Texas: Business Publications, 1979).

Gordon, Myron J., and Gordon Shillinglaw, *Accounting: A Management Approach* (Homewood, Illinois: Richard D. Irwin, 1974).

Horngren, Charles T., *Cost Accounting: A Managerial Emphasis* (Englewood Cliffs, New Jersey: Prentice-Hall, 1977).

Kieso, Donald E., and Jerry J. Weygandt, *Intermediate Accounting* (New York: John Wiley & Sons, 1977).

Kieso, Donald E., and Jerry J. Weygandt, *Intermediate Accounting* (New York: John Wiley & Sons, 1980).

Meigs, Walter B., A. N. Mosich, and E. John Larson, *Modern Advanced Accounting* (New York: McGraw-Hill Book Company, 1979).

Montgomery, A. Thompson, *Managerial Accounting Information* (Menlo Park, California: Addison-Wesley Publishing Company, 1979).

"Professional Notes," *The Journal of Accountancy,* December 1979.

Pyle, William W., John Arch White, and Kermit D. Larson, *Fundamental Accounting Principles* (Homewood, Illinois: Richard D. Irwin, 1978).

Rayburn, L. Gayle, *Principles of Cost Accounting with Managerial Applications* (Homewood, Illinois: Richard D. Irwin, 1979).

Welsch, Glenn A., Charles T. Zlatkovich, and Walter T. Harrison, Jr., *Intermediate Accounting* (Homewood, Illinois: Richard D. Irwin, 1979).

Weston, J. Fred, and Eugene F. Brigham, *Managerial Finance* (Hinsdale, Illinois: The Dryden Press, 1978).

CHAPTER 10

MANAGEMENT REPORTS

In addition to the reports recommended for general use in Chapter 9, some reports must be prepared specifically for management use. These reports include income statements, specialized project reports, profit plans, and forecasts.

INCOME STATEMENTS

From the accounting perspective, the going concern concept emphasizes the importance of the income statement. From management's point of view, the use of assets to produce net income is paramount. The income statement data must be prepared in a form that is most useful to the construction firm's management. The income statement, normally prepared on a monthly basis, must include comparisons with the preceding month, the same fiscal month last year, and the budget or projection. In addition, the statement must be presented in common terms, that is, the percentage of each section to sales and prior periods must be calculated.

The income statement (for management) may be prepared on a contribution or a fully absorbed basis. The fully absorbed income statement is usually prepared in a conventional fashion. Cost of sales, including all construction overhead (except home office) is deducted from total project revenue to calculate gross margin. Period expenses (home office overhead) are deducted from gross margin to arrive at net income before income taxes. The contribution income statement does not include allocated overhead in cost of sales. Allocated overhead is deducted as a total or period cost as is home office overhead. Instead of gross margin, the result is contribution margin, and after period costs the result is net income before income taxes equal to the fully absorbed statement.[1] Exhibit 18 includes a contribution and a fully absorbed income statement.

Whether prepared on a contribution or a fully absorbed basis, the management income statement must be supported by the performance reports for each construction project. If the firm's managers wish to review the details for the income statement, the project reports will add to the income statement. The results reported in the income statement *are* the addition of the project performance reports.

PROJECT REPORTS

The project performance reports that are presented to the firm's managers are substantially the same as those for external reporting and other internal uses except for percentage calculations and comparative information. The comparative information presented in the management reports should be by project, project phases, and details of project phases. The various levels of reporting should be additive and can be grouped by project or customer if desired.

The details should be analyzed to make the information more useful to management. The comparative data by bid or budget should be represented as an amount and as a percentage. The actual results must be expressed as a percentage of the objective (bid or budget) as well as the amount of difference. The amount and percentage change from last year should be included. The percentage of completion of each project and project phase is included for managers to evaluate the performance for the period reported. It is also necessary to include the basis for the calculation of the percentage completion. An example of a project report that includes this information is presented as Exhibit 19.

PROFIT PLANNING

Profit planning is very difficult in the construction industry. Any profit plan is affected by the randomness of the receipt of requests for proposals, the randomness of contract awards, the weather, and strikes. One method of estimating the possible awards is experience. Based on experience with past awards, the percentage of awards

[1] Ray H. Garrison, *Managerial Accounting, Concepts For Planning, Control, Decision Making* (Dallas, Texas: Business Publications, 1979), pp. 140-141.

from bids submitted is calculated. If the firm also sells a construction product, the trend line can be added to the contract expectations.

Another method of estimating is by statistics. The bids submitted are analyzed to find those that appear most likely to be awarded to the firm because of its expertise. The formula can be applied to the remaining bids to compute the possible awards. These estimated awards are the basis for the profit plan. This approach is more efficient if a computer is available for making the calculations.

Profit plans are necessary for the firm's fiscal year and for the fiscal month. If the information is not available for a profit plan for the fiscal year, a plan must be prepared for a minimum of the next fiscal month. Regardless of how crude the monthly forecast may be, the firm must look ahead at least one month.

THE NEED FOR PROFIT PLANS

In addition to the management's need for a forecast, there are many other outside requirements. A profit plan may be necessary to raise funds. Bankers, bondholders, and other lenders not only want to review history, but also like to know where the firm is going. Although creditors would like to look at least a year into the future, they will usually accept whatever is available.

The construction firm may find itself excluded from certain markets without a profit plan. Many jurisdictions (cities, counties, towns, townships, states, and federal agencies) require prequalification before the firm is allowed to bid within the jurisdiction for projects awarded by that jurisdiction. In some cases the prequalification requires historical information and forecasts or profit plans. Often the projections required for prequalification must cover a minimum of one year.

TOP DOWN PLANS

Although the discipline of management covers the alternatives of centralization and decentralization and the degree of participation, the nature of the construction industry may require the use of top down profit plans. The president of the firm and his advisors may be required to provide the estimates of future performance based on their knowledge, experience, and/or desires. Participation by part or all of the organization may be too time consuming and costly to be practical and may not lead to more useful information.

The firm's top management may use history adjusted for inflation to project profits for the coming months or year. Last year's performance can be adjusted for a percentage of increase or decrease based on expected contract awards. The results of this projection are then adjusted for inflationary expectations. The inflation adjustment should be applied to each cost category individually. The categories of cost involved, the management preferences, and the industry in which the firm does most of its construction projects determine the particular inflation adjustment to be used. The alternatives available include the Consumer Price Index, the Wholesale Price Index, or the Gross National Product Implicit Price Deflater.

Any profit plan must, to be effective, be compared to the actual results. The comparison must include a breakdown by category, such as direct materials, direct labor, overhead (in each category), and revenue. The object of the comparison is to identify the variance from the profit plan by each category. After the variance has been identified the causes must be determined. Variance analysis answers the questions of why and what can be done in the future. Analysis by cost category facilitates the identification of causes.[2]

FORECASTING

The profit plan is based on the forecast of costs and revenues. Construction project direct material can be estimated from bids made adjusted for those that are not expected to be awarded. Contracts awarded and bids that are probable contract awards contain the details of the direct material required and are the basis for the direct materials forecast. To these amounts should be added a contingency for unexpected awards and inflation.

For construction project direct labor, the hours by type of direct labor are summarized from expected and actual contract awards. The hours are then multiplied by the current direct labor rates for each labor category adjusted for inflation. The construction project direct labor hours projected by labor category are compared to the direct labor hours available by category to verify that the hours of direct labor needed are available.

SUBCONTRACT AND OVERHEAD FORECASTS

Subcontracting requirements are estimated from the contracts and possible awards by analyzing the bids and contracts. A construction firm usually solicits proposals from its subcontractors before sub-

---

[2]Charles T. Horngren, *Cost Accounting: A Managerial Emphasis* (Englewood Cliffs, New Jersey: Prentice-Hall, 1977), pp. 129-131.

mitting bids in response to a request for proposal. Even though these proposals from the subcontractors are available for forecasting, the construction firm is not bound, in an award, to use the subcontractors. The construction firm compares the costs of doing all of the project work internally with the costs of subcontracting (in areas in which the firm has the expertise available internally). The firm forecasts and uses the method that results in the higher net income. The forecast must, again, include a contingency to provide for inflation and any unexpected events or costs.

For a complete forecast, overhead by project and for the total organization must be estimated. Overhead costs that are peculiar to each project are estimated from the bids and contracts adjusted for any contingencies or inflation. Overhead for the total organization is estimated from the organization budgets or from departmental forecasts. These forecasts must also include any adjustments for inflation or contingencies.

FORECASTING HAULING AND EQUIPMENT USE

Forecasting equipment needs and the costs of hauling require a detailed analysis of the bids submitted and the contracts awarded. Hauling costs are estimated from site preparation and resurfacing. Equipment use is calculated from the anticipated projects. The estimated hours of equipment use are extended by the estimated cost rate. In addition, the hours required must be compared to the equipment hours available reduced by the estimated equipment downtime. It is also necessary to compare the types of equipment available with the types of equipment required. Management must know if other types of equipment are needed. Then they can decide whether it is economical to rent, lease, or purchase the needed equipment.

REVENUE AND CASH

Revenue and cash are forecast from the contracts in house and anticipated awards. The revenue may be forecast on a monthly basis, but the cash must be forecast at a minimum of weekly intervals. The receipt of operating cash can be anticipated from the projected revenue. The cash inflow is estimated from the payments of invoices by customers. The billing and accruals create the revenues and the payments cause the cash inflow. Nonoperating cash is generated from the sale of assets, collection of rents, and proceeds from loans. The cash outflow is estimated from the project forecasts and deducted from the inflow. The result is the net cash inflow by week (or month).

An important point to remember is the need for the forecasts to be prepared on time. Any delay to increase the accuracy of the forecasts, which makes the forecast less useful to management, should be avoided. Some degree of accuracy in the forecast should be surrendered to get the forecasts to management for use on a timely basis.

SUMMARY

Chapter 10 analyzes the income statement for management. The analysis includes the accounting concept of the going concern and the need for internal comparative data. The need to pyramid the income statment from the project performance reports is emphasized. Income statements prepared on a fully absorbed or on a contribution basis are described.

Information needed by management that should be included in project performance reports is discussed. This information includes percentages of completion and percentages of change from comparative data. Also included is a comparison of data by phase and of details within phase.

Management's need for profit planning is introduced in this chapter. The difficulty of profit planning in a construction environment is pointed out. Profit planning intervals are suggested. The possible use of top down rather than participatory estimates is examined. And the use of variance and variance analysis is described.

Forecasting methods are illustrated in this chapter. The descriptions include forecasting construction project direct material, direct labor, subcontracting, hauling, equipment use, and overhead. Total organizational overhead forecasting is treated separately.

The need for weekly cash flow information is affirmed. Again, emphasis is put on the requirement that forecasts and profit plans be timely to be useful!

EXHIBIT 18. Income Statements

Description	Amounts (In thousands)
Fully Absorbed	
Revenue (from contract billings and accruals)	$100,400
Cost of sales	
Project direct materials	$ 20,000
Project direct labor	$ 10,000
Project subcontracting	5,000
Project hauling	1,000
Project equipment use	1,000
Project direct overhead	2,000
Project allocated overhead	2,000
Total cost of sales	$ 41,000
Gross margin	$ 59,400
Period costs	
General and administrative overhead (home office)	$ 9,400
Net income before income taxes	$ 50,000
Contribution Basis	
Revenue (from contract billings and accruals)	$100,400
Variable cost of sales	
Project direct materials	$ 20,000
Project direct labor	10,000
Project subcontracting	5,000
Project hauling	1,000
Project equipment use	1,000
Project direct overhead	2,000
Total variable cost of sales	$ 39,000
Contribution margin	$ 61,400
Period costs	
Project allocated overhead	$ 2,000
General and administrative overhead (home office)	9,000
Net income before income taxes	$ 50,000

NOTE:
Either of these statements may be prepared for any desired time interval.

EXHIBIT 19. Project Performance Report for the Month Ended July 19XX

Description	Actual Amount	Forecast Amount	Difference Amount	Difference Percentage
		(In thousands)		
Project #8004				
Phase I				
Direct materials	$ 3,000	$ 2,500	$ 500	+20.0
Direct labor	1,000	900	100	+11.1
Subcontracting	1,000	800	200	+25.0
Hauling	100	85	15	+17.6
Equipment use	200	190	10	+5.3
Direct overhead	1,200	1,000	200	+20.0
Allocated overhead	1,200	1,100	100	+9.1
Home office overhead	1,100	900	200	+22.2
Total cost of phase I	$ 8,800	$ 7,475	$1,325	+17.7
Phase I revenue	$16,000	$15,000	$1,000	+6.7
Phase I net income	$ 7,000	$ 7,525	$ (325)	−4.3
Total expected revenue from phase I	$32,000			
Total percentage completion for phase I	40%			
Phase II				
Direct materials	$ 1,500	$ 1,250	$ 250	+20.0
Direct labor	$ 500	$ 450	$ 50	+11.1
Subcontracting	500	400	100	+25.0
Hauling	100	80	20	+25.0

EXHIBIT 19. (Continued)

Description	Actual Amount	Forecast Amount	Difference Amount	Difference Percentage
		(In thousands)		
Equipment use	100	100	-0-	-0-
Direct overhead	600	500	100	+20.0
Allocated overhead	600	500	100	+20.0
Home office overhead	600	400	200	+50.0
Total cost of phase II	$ 4,500	$ 3,680	$ 820	+22.3
Phase II revenue	$ 8,000	$ 7,500	$ 500	+6.7
Phase II net income	$ 3,500	$ 3,820	(320)	−8.4
Total expected revenues from phase II	$16,000			
Total percentage completion for phase II	39%			
Phase III				
Direct materials	$ 9,000	$ 8,000	$1,000	+12.5
Direct labor	3,000	3,000	-0-	-0-
Subcontracting	3,000	2,000	1,000	+50.0
Hauling	300	200	100	+50.0
Equipment use	600	500	100	+20.0
Direct overhead	4,000	3,500	500	+14.3
Allocated overhead	4,000	2,000	2,000	+100.0
Home office overhead	2,000	1,000	1,000	+50.0
Total cost of phase III	$25,900	$20,000	$5,700	+28.2
Phase III revenue	$50,000	$45,000	$5,000	+11.1

EXHIBIT 19. (Continued)

Description	Actual Amount	Forecast Amount	Difference Amount	Difference Percentage
			(In thousands)	
Phase III net income	$24,100	$24,800	$ (700)	−2.8
Total expected revenue from phase III	$100,000			
Total percentage completion for phase III	38%			
Total Project				
Direct materials	$13,500	$11,750	$1,750	+14.9
Direct labor	4,500	4,350	150	+3.4
Subcontracting	4,500	3,200	1,300	+40.6
Hauling	500	365	135	+37.0
Equipment use	900	790	110	+13.9
Direct overhead	5,800	5,000	800	+16.0
Allocated overhead	5,800	3,600	2,200	+61.1
Home office overhead	3,700	2,300	1,400	+60.9
Total project cost	$39,200	$31,355	$7,845	+25.0
Total project revenue	$74,000	$67,500	$6,500	+9.6
Project net income before income taxes	$34,800	$36,145	($1,345)	−3.7
Total expected project revenue	$148,000			
Total project estimated physical percentage completion	39%			

NOTE:
This performance report may be prepared for any desired time interval, such as a month or a year.

QUESTIONS

1. Describe the accounting concept of a going concern.
2. What type of information is considered comparative data?
3. What type of reports should support the income statement?
4. Describe the two alternative bases for preparing an income statement.
5. How should the details of a project report be presented?
6. Should a project performance report be disaggregated?
7. What makes profit planning difficult in the construction industry?
8. Is profit planning needed in the construction industry? If so, why?
9. Discuss the use of "top down estimates" for forecasting.
10. What are variances and how should they be reported?
11. Describe the forecasting of project direct material.
12. How is direct labor forecast?
13. How is subcontracting, hauling, and equipment usage forecast?
14. Describe the forecasting of project and home office overhead.
15. Is a cash flow forecast needed by a construction firm's managers?
16. Discuss the need for timely reporting.

REFERENCES

Anthony, Robert N., and James S. Reece, *Management Accounting* (Homewood, Illinois: Richard D. Irwin, 1975).

Chorba, George J., *Accounting For Managers* (New York: American Management Association Extension Institute, 1978).

Coombs, William E., and William J. Palmer, *Construction Accounting and Financial Management* (New York: McGraw-Hill Book Company, 1977).

Dearden, John, *Cost Accounting and Financial Control Systems* (Reading, Massachusetts: Addison-Wesley Publishing Co., 1973).

Garrison, Ray H., *Managerial Accounting, Concepts for Planning, Control, Decision Making* (Dallas, Texas: Business Publications, 1979).

Gordon, Myron J., and Gordon Shillinglaw, *Accounting: A Management Approach* (Homewood, Illinois: Richard D. Irwin, 1974).

Horngren, Charles T., *Cost Accounting: A Managerial Emphasis* (Englewood Cliffs, New Jersey: Prentice-Hall, 1977).

Kieso, Donald E., and Jerry J. Weygandt, *Intermediate Accounting* (New York: John Wiley & Sons, 1977).

Kieso, Donald E., and Jerry J. Weygandt, *Intermediate Accounting* (New York: John Wiley & Sons, 1980).

Meigs, Walter B., A. N. Mosich, and E. John Larson, *Modern Advanced Accounting* (New York: McGraw-Hill Book Company, 1979).

Montgomery, A. Thompson, *Managerial Accounting Information* (Menlo Park, California: Addison-Wesley Publishing Company, 1979).

"Professional Notes," *The Journal of Accountancy*, December 1979.

Pyle, William W., John Arch White, and Kermit D. Larsen, *Fundamental Accounting Principles* (Homewood, Illinois: Richard D. Irwin, 1978).

Rayburn, L. Gayle, *Principles of Cost Accounting with Managerial Applications* (Homewood, Illinois: Richard D. Irwin, 1979).

Welsch, Glenn A., Charles T. Zlatkovich, and Walter T. Harrison, Jr., *Intermediate Accounting* (Homewood, Illinois: Richard D. Irwin, 1979).

Weston, J. Fred, and Eugene F. Brigham, *Managerial Finance* (Hinsdale, Illinois: The Dryden Press, 1978).

CHAPTER 11

MANAGEMENT INFORMATION SYSTEMS (DATA PROCESSING) AND BONDING

The choice of data processing systems involves more than the selection of hardware. Applications, programming, and personnel must be selected as carefully as the hardware. In this chapter the selection and types of typical data processing applications in the construction industry are discussed. In addition, because of the nature of the industry, bonding and mechanic's liens are reviewed.

COMPUTER HARDWARE

The computer hardware market today has a large number of computers and peripheral equipment. They differ in size, speed, and ease of programming. One of the most important decisions in the choice of computer hardware is whether the data processing function is to be centralized or decentralized. The factors that influence this decision are the need for flexibility and less control versus the need for review and control of the use of management and financial information. Decentralization of computer hardware can be expensive, but so can the lack of information needed for a decision.

For the construction industry centralization is probably the best policy. The fact that most construction work is contracted with the need for legal review and the allocation of resources supports the requirement of centralization of at least the selection and use, if not physical control, of all data processing hardware. One of the objectives of centralization is to match the computer hardware size with the informational needs of the organization.

Bids should be requested from the manufacture of the type of hardware required. The bids must be evaluated centrally by both the financial division and the data processing group. If this expertise is not available to the construction firm internally, it can be purchased from independent CPAs and/or consultants. The analysis of computer hardware applies to central processors used for financial and management information as well as those used for scheduling or to simulate and compute structural and other civil engineering information. If possible, before deciding visit a construction firm with a similar computer.

Choosing peripheral equipment is a function of the data processing environment. One determinant is the need for output information and the form the output will take. In addition, the equipment needs will vary depending on the degree of decentralization of data processing and on the presence of a data base system.

Input hardware is chosen based on the form of input documents. If cards are needed for input or the environment is too primitive for tape or disc equipment, card punches and card to tape equipment is required. If the circumstances permit, key to disc or key to tape equipment is more efficient because it eliminates many additional operations, such as conversions.

A data processing or information systems installation requires internal security and environmental controls. Security within the system is needed to prevent the compromise of cash and other assets. Also, the system security must prevent each system from having access to all other systems. Each system must have access to only those other systems that are necessary to meet its objectives. For example, the accounts receivable system and users of that system must be prevented from accessing and using the payroll system, thereby preventing unauthorized use of payroll information. When a system accesses another system, the design must permit only the authorized use of the information accessed.[1]

PROGRAMMING

Programming is also known as software. The technical difference is that programming is the instruction in machine language that tells the machine what operations to perform on which data. Software refers to all the procedures involved in making the machines operate, such as system studies, programming, and system design.

[1]Robert G. Murdick, Thomas C. Fuller, Joel E. Ross, and Frank J. Winnermark, *Accounting Information Systems* (Englewood Cliffs, New Jersey: Prentice-Hall, 1978), 371, 383-384.

Software is equally important as hardware to the success of the data processing installation. The hardware cannot be used to the maximum extent possible without the proper software. In addition, the improper use of hardware can be caused by faulty software.

INTERNAL VERSUS EXTERNAL SOFTWARE

It is not necessary to develop all software internally, nor to purchase all software externally. A successful data processing operation can use a combination. To develop software the firm must have the necessary expertise "in house." Hiring the necessary experts can be expensive in both the short and the long run depending on the number of systems to be developed. An advantage of hiring internal experts is that they become familiar with the firm's operations and can tailor the software to the firm's particular needs. However, there will be a period of learning about the firm or of learning about software if the firm's personnel are to be trained in data processing.

Purchasing software has both advantages and disadvantages. One of the advantages is that the software has usually been tested through use and application. The prospective purchaser can check references and visit installations and observe the results of use of the software. Also, the purchase of software may be less costly to the firm in the long run. A disadvantage is that the software may not be exactly or most efficiently suitable for the firm's applications. Modifications may be expensive. Although it may be available faster than internal development of a firm's own software, it may also be more expensive in the short run and usually will, in the short run, require more cash outflow.

HIRING PROGRAMMERS

The hiring of programmers is a function of the size of the installation. If the installation is small, the security requirement for separation of duties cannot be met. In that case the firm should seek the service of generalists. For larger installations the programming should be separated into functions. New systems programmers should be employed for new systems applications. This type of programming requires originality and an ability to plan and control progress.

Maintenance programmers are responsible for programming updates of systems that are already in operation. Maintenance programmers must be original, but they must also be able to recognize the need for change and the application of the latest programming techniques to existing programs.

Operations programmers are needed to process systems as efficiently as possible. The operational programmers modify programs to take the utmost advantage of the capabilities of the hardware. They modify system's programs to allow the system to process efficiently on the firm's hardware.

Programming may be purchased by the hour or through a contract that specifies results. From the point of view of cost this alternative often appears to be the most attractive. However, the knowledge of the firm's operations may not be available. There is also the possibility that the firm may not be able to control the programming schedule. Any purchase agreement must include incentives or penalties for meeting or not meeting completion dates.[2]

SYSTEMS ANALYSTS

Systems analysts are the most sought after and expensive of the data processing specialists. Their function is to define, flow chart, and design systems. When the system analyst is through, the system should be ready for programming. Systems analysts should be familiar with and participate in the planning and control of their projects.

Systems analyses, like programming, is separated into new and maintenance systems analysis. The new systems analysts define and develop new systems, that is, systems that were not available in the firm before. The maintenance systems analysts update and maintain systems already in existence. Maintenance analysts are usually already familiar with the system or systems that they are maintaining.

It is possible but not feasible to purchase systems analysis. If systems analysis is purchased as part of a total purchase of data processing services, the seller's employees can develop some familiarity with the construction firm's systems and operations. Usually, however, the need for an intimate knowledge of the construction firm's operations makes the purchase of systems analyses services both expensive and difficult to find.[3]

MANAGING COMPUTER INSTALLATIONS

Computer installations must be managed as well as any other part of the firm. Competent computer specialists are not necessarily good

---

[2]Ibid., pp. 383-465.

[3]*Ibid*., pp. 248-286.

managers. In addition, because of the lack of knowledge about computer installations, the construction firm's other managers tend to leave the management of the installation to specialists. The result is often a grossly mismanaged management information systems department.

CONTROLLING THE COMPUTER

The computer installation must be viewed as another resource. This resource must be managed like shop equipment, buildings, or an assembly line and used intensively, with responsibility assigned for optimum results. The computer professionals are experts in information processing—not management! The construction firm's line management must impose controls and objectives on the computer installation.

One method for controlling the computer installation is to use zero base budgeting. With this method the computer installation must charge all or most of its costs to user departments. At the end of the accounting period its departmental cost must be at or near zero. This forces the computer installation to "sell" its services to other departments. The computer installation (data processing or management information systems) needs to compete with outside service bureaus and with other methods of processing data, including manual methods. Forcing data processing to seek the user tends to make it more efficient and responsive to the needs of the construction firm.[4] Because of the unique requirements of the construction firm, effective zero based budgeting may not be useful for all of the services of data processing. However, if zero based budgeting is used with the performance measures suggested below, the probability is that data processing will be more effective and efficient. An example of a zero based budgeting report for data processing is included in Exhibit 20.[5]

PERFORMANCE EVALUATIONS

A number of performance evaluations can be applied to data processing. Input performance can be measured by the number of correct input strokes per job, hour, or other interval. The number of strokes less the number of errors can be divided by a standard to create an input index. The effectiveness of input operators can then be numerically compared. A rational evaluation of the employees

[4]Charles H. Mott, *Accounting Reports For Management* (Englewood Cliffs, New Jersey: Prentice-Hall, 1979), pp. 133-135.

[5]*Ibid.*, pp. 136-137.

and the management of the data processing input section is the result.

For programming, the programming objective should be expressed in hours. The hours can be determined from industry and functional standards, from judgment, or from both. The actual hours used are compared to the objective by program and the variance is evaluated.

Systems analyses should be measured project by project. Before a systems analysis is started, the cost, the hours, and a tentative system completion date must be determined. Progress is measured periodically (monthly or weekly) against the objectives. If there is a material deviation from the plan, the objectives must be revised and the reasons for the deviation investigated. The systems analysis group, which usually participates in the setting of objectives, is responsible for meeting its objectives.

Computer hardware is used for differing hours from firm to firm. Some construction firms use their computer 24 hours a day every day of the week (with an allowance for downtime). The construction firm's line managers must choose the number of hours they estimate the computer will be used and then staff accordingly. When the computer is ready for operation, the running time must be reported, with a comparison to the estimated or available hours less downtime. This is an important measure for evaluating the hardware, the service from the supplier, and the effectiveness of computer applications. If the application objectives are not being met, the solution may be an increase in running time.

The performance of computer operators is important to the success of a computer installation. To prevent compromise and for the security of private information, the operators of the computer should work from detailed instructions that do not disclose or describe the nature of the program or system. The operator should be able to run any system without knowledge of the system or the program content. One measure of the effectiveness of the computer operators is the actual running time for a system or program versus the estimated time. Problem reports should be prepared to explain the variance, with the reasons and corrective action indicated. These comparisons must allow for the level of training and knowledge of the operator.

An additional measure is the job rerun report. If a job (system or program) must be rerun, it is an indication of problems and of the waste of computer and operator time. System and program reruns must be thoroughly investigated. The causes must be corrected.[6]

[6]Murdick, *op. cit.*, pp. 371-401.

As mentioned above, the total data processing function can be subcontracted. The subcontractor in these circumstances will do the complete function. Subcontracting costs must be analyzed and compared with other alternatives. The construction firm must include the cost of at least one employee to act as liaison with the subcontractor. Again, the dangers of subcontracting are the lack of knowledge of the construction firm's operations, the need for security of the construction firm's information, and the commitment of the subcontractor to the processing and reporting dates.

The decision to purchase all or part of data processing services and to choose the nature and size of the computer requires the use of the best available analytical techniques. The decisions of lease or buy and remaining with the manual system are part of the alternatives. The appropriate analytical techniques include the projection of cash outflows and inflows from the various alternatives and discounting the net inflows to present value. Other useful techniques include the discounted payback period and the discounted cash flow return.

COMPUTER APPLICATIONS

The applications of computers to management systems in the construction industry are slightly different from those in other industries. The numerous unions, the uniqueness of the contracts, and the diverse geographic locations cause differences in applications. The applications listed below are examples of the use of computers in the construction industry:

(*a*) Payrolls—Payrolls include the truck hire discussed in Chapter 9. The payrolls for officers, salary employees, and hourly employees are calculated and the checks printed on the computer. The information for journal entries and tax reporting are by-products of this system. Either a communication device or a delivery service must be used for remote payrolls, that is, projects in remote locations or long distances away from the computer operation.

(*b*) Project Costing—Project costing is a must for a computer installation. The need for timely management information requires the use of the computer for project performance reporting. Because of the mass of data and the need for accuracy, the computer is the most efficient method for reporting, recording, and storing project information.

(*c*) Estimating and Bidding—Responding to requests for proposals is one of the most important functions in the construction field. The bid is the source of sales or project revenue. Storing key bid information and the fast and accurate preparation of bid documents are important for the success of the construction firm. The rush of bid submission can be avoided with the computer. The more time can be allowed for review of the bid, the less the probability of a costly error.

(*d*) Scheduling—The computer can not only make scheduling easier, but it can make it possible! Scheduling is necessary for the use of construction equipment and project and phase completion, as well as for adjustments caused by nearing the end of the construction season or by unanticipated delays. The computer processes a mass of complicated data quickly and efficiently.

(*e*) Equipment Analyses—Buy or lease decisions are necessary for the acquisition of construction equipment. The computer can make the large number of calculations quickly and accurately.

(*f*) The General Ledger—A computer general ledger is neat and accurate and can be produced very quickly. And computer general ledger software is available from many equipment manufacturers as well as from software distributors.

(*g*) Financial Statement Preparation—All of the required financial statements can be programmed on the computer. They are the natural output of the general ledger and the incremental cost of including them is small. External, internal, contribution, or full absorption statements may be produced.

(*h*) Accounts Receivable—Because of the large amount of data, the accounts receivable is a suitable computer application. Part of the accounts receivable program can be a cash inflow report. The collection experience with each customer can be the basis for a projection of cash inflow.

(*i*) Credit Analyses—Once the accounts receivable have been captured on the computer and an order entry system installed, the computer can prepare and analyze customer credit information. In many parts of the construction industry, this information is basic and may be less important than other applications.

(*j*) Financial Analyses—Financial analyses for decisions is an important part of computer applications. The calculation of discounted cash flow return, present value, and payback are well suited for the computer. The computer can manipulate the large quantities of calculations required speedily and accurately.

BONDING

Bonding is necessary in the construction industry. The unique nature of the product and the heavy reliance on formal contracts require the protection of the buyer and the supplier through performance guarantees. The types of bonds required depend on the nature of the guarantee.

BID BONDS

As part of the bid submission in most jurisdictions, the bidder must include evidence of bonding. The requirement of bonding of bidders is thought to be an important part of prequalification of construction firms because the inability to provide guarantees prevents the construction firm from prequalifying. The reasoning is that only responsible bidders can meet the requirements for bid bonding.

The bid bond protects the owner (general contractor, customer) from either nonperformance on the part of the successful bidder or the failure of the successful bidder to qualify for a performance bond (to be discussed below). Should the successful bidder not sign or perform the contract, the bonding company (surety) will indemnify the customer for the difference between the bid of the construction firm chosen and that of the next lowest responsible bidder. The indemnification, however, is limited to the maximum bid received or calculated or the bond penalty amount (the amount of liability identified in the bid bond).

Rarely will the bonding company reimburse the customer or the organization issuing the contract. Usually the bonding firm will find another construction firm to complete the contract or will let the construction firm awarded the contract "buy back" the bid and perform the work in a different relationship to the owner or customer. The bonding company will try to avoid this possibility through screening and review of its customers and their financial status.

Should the bonding company be required to perform or indemnify the customer under the terms of its bond, the bonder has recourse to the assets of the bondee (the construction firm or customer of the bonding company). In theory, at least, it makes sense for the bonding company to verify its customers' ability to reimburse it for any loss before issuing (selling) the bid bond to the construction firm (the bonding company's customer).

[7]J. Peterson, "Bonding of Contractors—A Surety's Analysis," *The Journal of Commercial Bank Lending*, Vol. 58, July 1976, pp. 32-45.

Although this discussion concentrates on the use of bid bonds, which is the most common practice, a cash deposit can be used in place of a bid bond. In some jurisdictions, a cash deposit at and by a commercial bank (for which the bank will charge the construction firm a fee) equal to the bond limit will suffice. This alternative was more popular in the past, but because it limits the use of cash on deposit instead of relying on the reputation and resources of the bonding company, it is not as widely used as the bid bond.

PAYMENT BONDS

In addition to the owner or customer, the construction firm's employees, subcontractors, and suppliers must be protected. The purpose of the payment bond is to guarantee payment to the employees, subcontractors, and suppliers of the firm. If the construction firm cannot pay a valid claim by a supplier, subcontractor, or employee, the surety (bonding company) will pay the claim. Should the bonding company pay claims on behalf of the construction firm, the bonding company will have recourse to the assets of the construction firm. The payment bond protects the interested groups from the bankruptcy of the construction firm and from the owners' absconding with the construction firm's funds and not paying the groups that performed the work or furnished material.

A word of caution, however. The bonding firm will only be liable to the extent of its agreement in the bond. And a bond is required for each different job or project. Part of the proof of claim is relating the claim to the job or project and therefore to each bond. Additionally, it must be remembered that not all construction jobs are bondable jobs. The requirement for bonding is usually specified in the construction firm's contract with the owner (customer or general contractor).

Because of the risk to and the involvement of the bonding company, the surety (bonding company) may require that the construction firm be audited by an outside (independent) firm of CPAs, with copies of the audit report submitted to the surety. Usually the bonding company will require the construction firm to submit a monthly report of the number of contracts (by job or project) completed during the month and a listing of the projects (contracts) open at the end of the month. The report also has the status of each job, that is, the amount of the award and the amount spent through the report date. The construction firm must notify the bonding firm of any new awards, with a bond application and a copy of the award.[8]

[8]*Ibid.*

PERFORMANCE BONDS

The performance bond guarantees the performance of the work according to the contract terms. The performance bond protects the owner (general contractor) from the nonperformance of the project by the construction firm after the contract has been executed. If the construction firm does not perform after the contract has been signed by both parties (executed), the bonding company will reimburse the owner for the difference between the original contract and the amount of the award of the contract to another construction firm.

Again, the bonding company may find another firm to perform work, allow the original firm to "buy back" the contract, or indemnify the owner in cash. Because of the contractural relationship between the bonding company and the construction firm, the bonding company has recourse to the assets of the construction firm to recover for any payments made under the terms of the bond.[9]

EXECUTING BONDS

The execution of a claim under a bond is an expensive and time consuming process. The result can be extensive and expensive litigation. There is a widespread belief that executing on a bond of the types used in the construction industry will only benefit the attorneys. Concrete evidence to support this belief is not available. However, the reader must remember that a bid, payment, or performance bond is a legal instrument and subject to the legal traditions of our society.

Sureties (bonding companies) are limited in the degree of their liability. There is always a point beyond which the bonding company is not required to and will not go. The bond and/or the bonding contract specifies the limit of the bonding company's liability on any one bond. The financial strength of the construction firm, the size of the contract, and the risk evaluation are all factors to be considered in determining the degree of the bonding company's liability. The cost of the bond to the construction firm is a function of the evaluation of these factors.

In addition to the bonding of each job or project to the limit agreed upon or specified, there is a limit to the total bonding of any one construction firm. The limit should, in theory, prevent the construction firm from overcommitting itself because the lack of bonding protection would eliminate the firm from additional contract awards.

[9]*Ibid.*

Usually the surety (bonding company) computes the limit of the bonding coverage it will grant to a single client. The use of an additional bonding company by the construction firm is impractical because of the reporting and audit requirements and the possibility that the firm may nevertheless need to remain within a total limit. The limit of total bonding to any single construction firm is based on the bonding company's judgment and calculation of the ability of the construction firm to meet the obligations of all of its jobs or projects. Although this definition is highly dependent on the application of judgment to financial information, as a general rule the bonding company will issue bonds to the limit of 10 times the amount of the construction firm's working capital.

Bonds terminate upon the satisfactory completion of the contract by the construction firm. The question is when is a contract complete? Completion is usually the acceptance of the job by the customer. This may be a formal process requiring the agreement of both parties or the satisfaction of provisions in the original contract. Usually the completion requirements are defined to prevent the interpretation "at the owner's convenience." For example, the beneficial occupancy of a building by the owner would be interpreted as completion.[10]

MECHANIC'S LIENS

Mechanic's liens are designed to protect the construction firm from nonpayment by the owner or customer (general contractor). Because most construction is either the building or improvement of real property that is not owned by the construction firm, the problem of collecting the payments due is unique. The construction firm is unable to reclaim its inventory or effort. Therefore, the mechanic's lien was designed to allow the construction firm to place a claim against the real property to prevent its use or sale without satisfaction of the claim. The lien prevents the sale of the real property without first satisfying the claim. If there is a lien and the property is sold, it is sold subject to the lien and the buyer can become liable for the claim. Usually, if the real property is sold, at the time of sale part of the proceeds are used to satisfy the claim and the claimant will sign a release of lien. If the property is intended for use by the owner or customer, the lien is a legal claim and may be pursued accordingly.

A mechanic's lien must be executed to be established. The

[10] *Ibid.*

claimant, usually a construction firm, must file a claim against the property in conformance with the lien law, to prevent the sale of the property without recognizing the claim. To be enforceable the lien must conform to the lien laws of the various states (jurisdictions). The claim must be filed within the proscribed time period and with a court or legal authority with jurisdiction over titles to real property. Usually, there is a requirement that the lien specify the exact location of and description of the property liened.[11] This requirement affects the contract documents that describe the job. These documents must include sufficient information in the event it is necessary for the construction firm to lien the property.

The lien laws vary from state to state. Some examples of the lien laws and their differences follow. The reader should realize, however, that the mechanic's lien laws of the state in which the project is located must be studied for the exact requirements. One example is based on the mechanic's lien laws of the state of California. In California there is a contractor's report (a publication similar to a trade newspaper) that lists completed jobs or projects. Completed does not mean the end of the work on the project. It means the project has been accepted or beneficially occupied by the customer or owner. The publication of the completion date is notice to claimants (construction firms) that they have 30 days from the completion date to file mechanic's liens for unsatisfied claims. If the lien is not filed within the 30 day period, the claimants lien rights expire and the property may be used or sold free of any prior liens.[12]

Maryland, said to be the state where the mechanic's lien originated, has a different approach. The construction firm has 30 days from the date of the last work performed on that job to file a mechanic's lien. However, if within 30 days from the last work performed on the project the construction firm files an intent to lien for unsatisfied claims, the claimant is granted, under the mechanic's lien statute, an additional 60 days within which to file the lien. The lien (or the intent to lien) must be registered with the proper legal authority and must include a legal description of the job site. If these requirements are not met, the lien rights will expire and the property may be used or sold free of any prior liens.[13]

---

[11]R. Clark, "Owners of Accounts Receivable of Contractors Analyzed," *National Underwriters (Fire Edition)*, Vol. 23, July 4, 1969, p. 26.

[12]"California Supreme Court Upholds Mechanics Liens," *Air Conditioning and Refrigeration News*, Vol. 139, October 25, 1976, p. 23.

[13]J. Barfield, "Extending Credit in the Construction Industry," *Credit and Financial Management*, Vol. 80, November 1978, pp. 29-33.

Usually, if the claim is paid, the owner or customer will ask the construction firm to execute a release of lien that clears the title for the owner and allows title transfer free of any claims. Often, whether a lien is filed or not, the owner or customer will request a release of lien before or at the time of final payment to the construction firm. The release of lien is evidence of the surrender of any limitation to the owner's title by the construction firm.

PREQUALIFICATION

Many jurisdictions (state and local governments) require that construction firm's prequalify before they are allowed to bid for work contracted by the jurisdiction. State and local laws differ as to the exact requirements for prequalification and these laws must be complied with to qualify. However, some general characteristics are found in many of these statutes.

One requirement is a time limitation on filing. The construction firm's prequalification data must be on file a certain number of months or days before the contractor is allowed to bid for that jurisdiction's work. In some cases, the requirement is that the data must be filed within a certain time period (before a date certain) if the firm is to be allowed to bid during a specified bidding period. Any data submitted after the specified period will delay bid permission until a future bidding period. The purpose of the delay between filing and bidding is to allow review of the contractor's data.

Another characteristic of prequalification is the need to update the data after the first filing. The jurisdictions require that the original data be updated periodically according to state and local laws. It is the construction firm's responsibility to update the information as required. If the updating requirement is ignored, the firm may be prevented from bidding.

One characteristic that is not universal is the requirement that the construction firm's submission be verified. Verification is required by many jurisdictions, however. The verification requirements differ depending on local laws. Some require that the submission be audited by an independent CPA. If the construction firm is audited at the end of its fiscal year but the submission is made at a later date and includes updated material, the submission may have to be audited again and an audit certification furnished to the jurisdiction. In some jurisdictions a past audit suffices, provided that it is updated through a review by the construction firm's independent CPAs. A copy of the review report must be submitted to the jurisdiction. If the construction firm has not been audited, the chances are that the jurisdiction will require an audit certification with the filing.

SUMMARY

This chapter is a discussion of data processing, bonding, mechanic's liens, and prequalification. For data processing the discussion includes the selection and acquisition of hardware, programming, and systems analyses. Hardware selection is influenced by the need and degree of centralized data processing. The relation of hardware to software is reviewed. The discussion includes a review of the alternatives of purchasing software or developing it internally.

Comments include the hiring of programmers and systems analysts and organizing these functions. Managing the computer installation is also discussed. Techniques for measuring the performance of the computer installation are listed. The advantages and disadvantages of subcontracting the total data processing function are reviewed. In addition, a list of the typical applications of data processing in a construction environment is included.

In this chapter the unique construction industry requirement for bonding is analyzed. The analysis includes bid bonds, payment bonds, and performance bonds. The need for bonding is presented with an analysis of the relationship between the construction firm and the bonding company. Bond execution is reviewed, with comments on the limits of the bonding company's liability and the calculation of total bonding limits for the construction firm.

Mechanic's liens and the need for them are summarized in this chapter. The summary includes examples of the differences between the lien laws of various jurisdictions. Maryland and California are chosen as illustrations. The strict legal requirements of lien statutes are emphasized.

Prequalification is discussed. The jurisdictional requirements that are common to most jurisdictions are listed. The filing requirements and the updating requirements are reviewed. Additionally, the possibility of the need for an audit certification by independent CPAs is noted.

EXHIBIT 20. An Example of a Zero Based Budget for Data Processing

Description	19XX Budgeted Amount	Amount Charged to Users	Actual Amounts Expended
	(In thousands)		
Computer Operations			
Equipment rental	$ 750	$ 800	$ 750
Operators salaries	90	196	100
Overhead	40	-0-	50
Total	$ 880	$ 996	$ 900
New Systems Programming			
Programmers' salaries	$ 100	$ 196	$ 120
Overhead	50	-0-	50
Total	$ 150	$ 196	$ 170
Systems Analysts			
Analysts' salaries	$ 120	$ 256	$ 130
Overhead	60	-0-	70
Total	$ 180	$ 256	$ 200
Operational Programming			
Programmers' salaries	$ 30		$ 30
Overhead	15		20
Total	$ 45	N/A	$ 50
Maintenance Programming			
Programmers' salaries	$ 75	$ 176	$ 80
Overhead	37	-0-	30
Total	$ 112	$ 176	$ 110
Planning and Equipment Evaluation			
Salaries	$ 40		$ 40
Support costs	15		20
Total	$ 55	N/A	$ 60
Data Conversion			
Equipment rental	$ 12	$ 15	$ 14
Salaries	45	126	60
Overhead	25	-0-	30
Total	$ 82	$ 141	$ 104

EXHIBIT 20. (Continued)

Description	19XX Budgeted Amount	Amount Charged to Users	Actual Amounts Expended
	(In thousands)		
Data Processing General Management			
Salaries	$ 120	$	$ 130
Support costs	60		75
Total	$ 180	N/A	$ 205
Grand total	$1,684	$1,765	$1,799

NOTES:

(1) This schedule is a summary of a budget and a year's activity for a zero budgeted data processing operation in a manufacturing environment.

(2) For this illustration, fringe benefit costs are included in overhead and support costs rather than as part of labor costs. Also, overhead and support costs include purchase and/or rental of software and software packages.

(3) The budgeted amounts are also used as the basis for charging users. Users of data processing services are charged for actual hours used at the budgeted rates. Differences between amounts charged to users and actual costs are either absorbed by or accrue to data processing.

(4) In this case, the amounts charged to users of data processing services exceed budgeted amounts. This was caused by the use of more hours for projects than anticipated. The charges to users are allocated through direct departments and include overhead costs as well as costs for indirect departments. The excess charged ($81,000) would be a profit to data processing if the actual costs incurred did not exceed the budget.

(5) The actual costs exceeded by the budget by $115,000 and exceeded the amount charged to users by $34,000. Because the actual costs exceeded that charged to users, data processing must show a loss or an unfavorable variance from its zero budget. The unfavorable variance was caused by the use of more time than anticipated on projects and the attendant overtime and other costs.

QUESTIONS

1. What factors should be considered when choosing computer hardware?

2. How should peripheral equipment be chosen?

3. How is input hardware chosen?

4. Discuss the importance of the environment and security to a computer operation.

5.   What is "software"?

6.   Should software be developed internally or purchased from outside sources?

7.   Define new systems programming.

8.   Define maintenance programming.

9.   Define operational programming.

10. What is systems analysis?

11. Differentiate between new systems analyses and the analysis of existing systems.

12. What are the advantages and disadvantages of purchasing systems analysis?

13. Discuss the management of computer installations.

14. How can the performance of input groups be measured?

15. What techniques can be used to measure the performance of programmers?

16. Describe a technique that can be used to control the performance of systems analysts.

17. Are there methods that can be used to measure the performance of computer operations?

18. What are the advantages and disadvantages of purchasing a computer rather than subcontracting the data processing function?

19. List and describe some of the typical applications of the computer in the construction industry.

20. What is a bid bond?

21. Describe the use of payment bonds in the construction industry.

22. What are performance bonds and how are they used?

23. How are bonds executed and what problems may be encountered?

24. Describe and discuss mechanic's liens.

25. What is prequalification?

REFERENCES

Anthony, Robert N., and James S. Reece, *Management Accounting* (Homewood, Illinois: Richard D. Irwin, 1975).

Barfield, J., "Extending Credit in the Construction Industry," *Credit and Financial Management*, Vol. 80, November 1978.

"California Supreme Court Upholds Mechanics Liens," *Air Conditioning and Refrigeration News*, Vol. 139, October 25, 1976.

Chorba, George J., *Accounting for Managers* (New York: American Management Association Extension Institute, 1978).

Clark, R., "Owners of Accounts Receivable of Contractors Analyzed," *National Underwriters (Fire Edition)*, Vol. 23, July 4, 1969.

Coombs, William E., and William J. Palmer, *Construction Accounting and Financial Management* (New York: McGraw-Hill Book Company, 1977).

Dearden, John, *Cost Accounting and Financial Control Systems* (Reading, Massachusetts: Addison-Wesley Publishing Co., 1973).

Garrison, Ray H., *Managerial Accounting, Concepts for Planning, Control, Decision Making* (Dallas, Texas: Business Publications, 1979).

Gordon, Myron J., and Gordon Shillinglaw, *Accounting: A Management Approach* (Homewood, Illinois: Richard D. Irwin, 1974).

Horngren, Charles T., *Cost Accounting: A Managerial Emphasis* (Englewood Cliffs, New Jersey: Prentice-Hall, 1977).

Kieso, Donald E., and Jerry J. Weygandt, *Intermediate Accounting* (New York: John Wiley & Sons, 1977).

Kieso, Donald E., and Jerry J. Weygandt, *Intermediate Accounting* (New York: John Wiley & Sons, 1980).

Meigs, Walter B., A. N. Mosich, and E. John Larson, *Modern Advanced Accounting* (New York: McGraw-Hill Book Company, 1979).

Montgomery, A. Thompson, *Managerial Accounting Information* (Menlo Park, California: Addison-Wesley Publishing Company, 1979).

Mott, Charles H., *Accounting Reports for Management* (Englewood Cliffs, New Jersey: Prentice-Hall, 1979).

Murdick, Robert G., Thomas C. Fuller, Joel E. Ross, and Frank J. Winnermark, *Accounting Information Systems* (Englewood Cliffs, New Jersey: Prentice-Hall, 1978).

Peterson, J., "Bonding of Contractors—A Surety's Analysis," *The Journal of Commercial Bank Lending*, Vol. 58, July 1976.

"Professional Notes," *The Journal of Accountancy*, December 1979.

Pyle, William W., John Arch White, and Kermit D. Larson, *Fundamental Accounting Principles* (Homewood, Illinois: Richard D. Irwin, 1978).

Rayburn, L. Gayle, *Principles of Cost Accounting with Managerial Applications* (Homewood, Illinois: Richard D. Irwin, 1979).

Welsch, Glenn A., Charles T. Zlatkovich, and Walter T. Harrison, Jr., *Intermediate Accounting* (Homewood, Illinois: Richard D. Irwin, 1979).

Weston, J. Fred, and Eugene F. Brigham, *Managerial Finance* (Hinsdale, Illinois: The Dryden Press, 1978).

CHAPTER 12

FINANCIAL ANALYSES

Financial analysis takes two forms—ratio analyses of financial statements and financial evaluations. Ratio analysis is used for internal and external measurement. Financial evaluations are used internally for making business decisions. The construction firm as well as other business firms use these analyses to measure performance. The major difference between the construction firm and other businesses is the greater degree of uncertainty inherent in decisions in the construction industry. Consequently, financial analysis assumes greater importance.

FINANCIAL RATIOS

Depending on the decision to be made, there are many more ratio calculations than those included in this chapter. However, the most commonly used are presented here. These ratios are particularly useful for general business evaluations of firms, including the construction firm.

THE WORKING CAPITAL RATIO

The working capital ratio is a general measure of the ability of a firm to liquidate its current liabilities. Without working capital to finance inventories, meet payrolls, and carry receivables the firm may not survive. The working capital ratio (also known as the current ratio) measures the firm's ability to liquidate its current liabilities and have enough working capital to continue operations.

To calculate the working capital ratio, the total current assets (the total is used and not the average because the values at a point in time are relevant to measuring the ability to liquidate liabilities and continue operations) are divided by the total current liabilities. The product or answer is the ratio of current assets to current liabilities expressed as a ratio such as 2 to 1 or 1.5 to 1, that is, there are $2 of current assets for $1 of current liabilities or $1.50 of current assets for $1 of current liabilities.

What is a good or acceptable current ratio? There is no simple answer to that question. The working capital ratio for any firm at any point in time should be compared to the historical ratio for the firm to identify a trend. Comparison with the last calculation indicates an improved or deteriorating status. Comparison with the historical trend indicates whether the firm is improving the ratio or not. In addition, the ratio can be compared with the ratio for the industry or the segment of the industry that is the firm's market. This information can be gotten from the trade association or from one of the investor services. There is not a universally "safe" ratio.

This ratio and the working capital amount are widely used by creditors (the working capital amount is the current assets minus the current liabilities and is expressed in dollars). It is not unusual for the working capital ratio required of the debtor by the creditor to be included in the credit (debt) agreement. A violation of the agreed upon ratio is a violation of the agreement and technically the loan becomes due and payable. The working capital ratio proposed by the creditor is usually based on either the firm's historical ratio or an industry standard. However, the proposed ratio is often negotiable and in the event of violation the creditor will temporarily waive the ratio requirement rather than call the loan.[1]

THE QUICK RATIO

This ratio is sometimes known as the acid test ratio. It is calculated by dividing the current assets that are easily or quickly turned into cash, such as cash, marketable securities, and accounts receivable, by the total current liabilities. As in the case of the current ratio, averages are not used. The amounts are those of the financial statements at ends of periods.

The quick ratio measures the ability of the firm to quickly extinguish its current liabilities through the use of current assets that can be immediately turned into cash. Inventory and other current assets that take longer to convert into cash are not included in the calculation. For example, inventory must be sold and then the

[1]Donald E. Kieso, and Jerry J. Weygandt, *Intermediate Accounting* (New York: John Wiley & Sons, 1980), p. 1180.

receivable collected within the terms allowed before it can be converted to cash. Therefore, inventory is excluded from the calculation of the quick or acid test ratio.

Creditors use this ratio as a measure of the firm's (including the construction firm's) liquidity. No standard quick ratio exists. Creditors will use the firm's historical average or an industry standard to evaluate the firm's status.[2]

RETURN ON COMMON STOCKHOLDER'S INVESTMENT

The return on common stockholder's investment is expressed as a percentage. It is a measure of the degree of earnings of the firm that are applicable to or in a sense belong to those who have put funds at risk in the firm. The ratio is calculated by dividing after tax net income less preferred stock dividends by the average common stockholder's equity. The average common stockholder's equity is calculated by averaging total stockholder's equity for the period and deducting from that average the average of the preferred stockholder's equity. The resulting percentage is the return to common stockholders for that period.

This ratio is widely used by investors. Although, as with the quick ratio, there is not a standard or universal ratio, the current period's ratio is usually compared with past ratios for the firm. The historical trend of the ratio is the best indicator. However, the firm is competing for funds within its industry and with all other firms that need capital. If the firm is to attract funds within its industry, its return on common stockholder's investment must compare favorably with the industry average or standard. The firm is also competing with all other firms in the capital market. Therefore, its return must compare favorably to attract capital.[3]

ASSET TURN

Asset turn is a measure of the relationship between net sales and the assets required to generate those sales. The asset turn answers the question of whether the firm is using its assets effectively or not! The turn is calculated by dividing the firm's average total assets into the firm's net sales for the period. The result is expressed as a ratio such as 1.35 turns or 0.95 turns. When compared to an average, the turn indicates whether too many assets are being used to generate sales.

The turn can also be calculated for individual assets such as

---

[2] *Ibid.*
[3] *Ibid.*, p. 1186.

inventory and accounts receivable. In the case of inventory, the firm's average inventory is divided into the net sales resulting in the number of times inventory was turned during the period to generate the sales. The average inventory is substituted for the average total assets. (Although it is more correct to divide average inventory into cost of sales, for comparative ratios that information may not be available. Consistency of method of calculation is equally important.) To calculate the accounts receivable turn, the average of accounts receivable is divided into the firm's net sales. The resulting turn can be divided into 360 or 365 to calculate the average days the receivables are outstanding. This average is compared to the firm's terms of sale to check for collection within terms. If the average number of days of receivables outstanding exceeds the sales terms, the collection effort and credit terms must be reexamined. The sales terms should not be exceeded by more than 10 days.

No standard turn exists for total assets, inventories, or accounts receivable. These measurements should be compared to the firm's historical turns for each asset classification. If available, the firm's turns should be compared to the industry average.[4]

MARGIN ON SALES

The percentage earnings on sales is an important measure of profitability. The percentage is calculated by dividing net income plus interest expense multiplied by 100% minus the effective income tax rate (expressed as a percentage, such as 20%) by net sales for the period. The result is a percentage, such as 11.5% margin on sales.

Again, as with most ratios, there is not a standard ratio. The ratio for the current period should be compared with the percentage for past years to identify any trends. The current percentage should also be compared to the average for the industry if that information is available.

This percentage can be calculated for net income both before and after income tax. The percentage described here is the after income tax calculation. The after income tax calculation is more useful because it measures the earnings percentage after all costs, including income taxes. A change in the percentage return from period to period, especially if unfavorable, may require changes in policy. The change could indicate a need for price increases, a need for an increase in sales incentives to increase sales, and/or the need for a cost reduction program to increase the net income earned on sales.[5]

[4]*Ibid.*, p. 1183.
[5]*Ibid.*, p. 1184.

RETURN ON CAPITAL EMPLOYED

This is a most important measurement. Expressed as a percentage, this calculation measures the return on the firm's total assets. The total assets represent the total investment in the firm by suppliers, shareholders, and other creditors. It is management's responsibility to earn the maximum possible on the total investment regardless or whether the investment came from shareholders or creditors.

Return on capital employed is calculated by dividing the average total assets into the firm's after income tax net income plus interest expense multiplied by 100% minus the effective income tax rate (expressed as a percentage, such as 20%), 100% − 20%. The results are expressed as a percentage return on total assets, such as 12.5%. The interest expense as adjusted is added back because it is part of the return earned for creditors. It is the peculiarity of our tax law that one form of earnings is deductible for income tax purposes.

To make this ratio (return) more useful for taking corrective action and for analysis, it can also be calculated by multiplying the asset turn by the return (margin) on sales. Should the total ratio change unfavorably, these ratios are more useful indicators of where corrective action is needed.

The comparative criteria for this percentage are historical returns and trends and industry averages. However, it must be remembered that this percentage (ratio) is a measure of management's ability to earn an adequate return on all the resources provided. Therefore, management is competing with all firms in the various capital markets. To successfully compete, the firm's return on total capital employed must compare favorably with the return for other firms.[6]

THE Z SCORE

This measure was developed to attempt to predict bankruptcies. The Z score is the result of the weighting of a number (5) of ratios. Although the trend in the Z score has meaning, as well as the industry score, each Z score calculation can be measured against a predetermined criterion. If the Z score is less than 1.81 (1.80 or less), the firm has a high probability (7 to 9 out of 10) of bankruptcy within two years. If the Z score is between 1.81 and 3.00, the firm has a random probability of bankruptcy within two years. And if the Z score is 3.01 or greater, the firm has a low probability (1 out of 10 or less) of bankruptcy within the next two years.

The first ratio in the Z score calculation is the period ending working capital divided by the average total assets for the period

[6]*Ibid.*, p. 1185.

multiplied by the factor of 1.2. For example (based on a large firm), the firm's period ending working capital of $731,000 divided by average total assets of $10,134,000 = 0.072 × 1.2 = the first input to the Z score of 0.09.

The second Z score ration is the average retained earnings for the period divided by the average total assets multiplied by the factor of 1.4. For example, the average retained earnings of $1,880,000 divided by average total assets of $10,134,000 = 0.186 × 1.4 = 0.26.

Period ending pretax net income plus interest expense for the period divided by the average total assets for the period multiplied by the factor of 3.3 is the third ratio in the Z score. Using the same example throughout, the pretax net income of $1,128,000 plus interest expense of $369,000 divided by average total assets of $10,134,000 = 0.148 × 3.3 = 0.49.

The fourth ratio in the calculation of the Z score is the division of the market value of the common stock plus the liquidating value of the preferred stock by the period ending long term debt. The market value of the common stock is calculated by averaging the four quarterly stock market sales price quotes. If the stock was not traded and no quotes are available, the book value per common share may be substituted. The liquidation value of the preferred shares is normally stated in the stock certificate. However, if this information is not available, the book value of the preferred shares may be used. An example of this calculation is: $298,000 market value of common stock + $264,000 liquidation value of preferred stock divided by $3,804,000 period ending long term debt = 0.148 × 0.6 = 0.09.

The fifth ratio is net sales for the period divided by the average total assets for the period, multiplied by the factor of 1.0. In this example the values are $58,762,000 in net sales divided by average total assets of $10,134,000 = 0.865 × 1.0 = 0.87.

Totaling the five components of the Z ratio results in a total score of 1.80. According to the criterion for evaluating the Z score, the total indicates a high probability of bankruptcy within two years. The example from which these calculations were made is a national, well known, large U.S. firm. The data are from the calendar and fiscal year 1979. As of the moment there are no indications that this firm is filing or will file for bankruptcy. The Z score indicates a high probability of bankruptcy to which this firm is the exception. Although the Z score may be used as another input and guide, the manager's and reviewer's judgment is necessary to evaluate all the available information about the firm.[7]

[7]"Bankruptcies—Foresee, Forestall, But Don't Fall Short," *The CPA Journal*, Vol. 49, No. 11, November 1979, p. 85.

INTERNAL USE OF RATIOS

Financial ratios are as important for use internally as they are for use by analysts and other external experts. Management is very interested in changes in financial ratios from period to period. Internal management (in construction as well as other industries) compares current period results with prior periods and with industry averages. Any changes are noted and investigated and corrective action is initiated. Internal managers react to unfavorable trends and attempt to reverse or correct the unfavorable conditions. These ratios (and others not included here) are calculated monthly or on an annualized or rolling basis. Internal managers want to calculate and compare ratios as soon as possible, that is, monthly or more frequently if possible. Therefore, corrective action can be initiated more quickly.

FINANCIAL EVALUATIONS

In addition to the use of ratios, financial evaluations are an important part of internal decisions. Integral to financial evaluation are the identification of the fixed (discretionary) and variable components of costs. The identification of cash inflows and outflows from each alternative and the use of present value or discounting are also necessary.

FIXED (DISCRETIONARY) AND VARIABLE COSTS

Financial evaluations require the identification of the fixed and variable components of all expenses and costs. However, this separation of costs does not have to be exact or have a confidence level of 100%. Many expenses do not fit neatly into either category of variability. Some expenses or costs need to be separated into that part which is variable and that part which is fixed. The high-low method or regression analysis may need to be applied to each expense account to separate the components.

Once the parts have been calculated, the fixed and variable factors should be applied to historical data to test their validity. If applying these factors to historical levels of production activity results in expense totals that are reasonably close to historical amounts, the disaggregation can be accepted. Should an expense test result in a material difference between the calculated value and the historical value, it is prudent to classify the expense as discretionary (fixed).

When calculating the fixed and variable factors, the traditional accounting records and reports are the basis for the calculation input. But the information required, the separation of cost variability, is usually not available in the accounting records. Therefore, the typical approach is to study the accounting records for the input data needed and update the study periodically, for example, every six months.[8] This information is not only useful for financial evaluations, but also for the submission of contribution bids during periods of slow activity. One factor that must be kept in mind when using variable and fixed data for internal decisions is the approximate relevant operating range of the construction firm. If the decision expands the firm's fixed investment, the expansion must be treated as a variable cost for that decision.[9]

For contribution bidding, the objective is to receive a cash inflow that exceeds the variable costs of the contract. Thus the firm will receive a contribution toward its fixed costs and profit. To submit this type of bid the firm must have an estimate of the breakdown of its costs into fixed and variable factors. Contribution bids are usually submitted during times of reduced construction activity when the firm's managers are willing to accept a contribution to fixed (discretionary) costs and keep the firm in business until the market improves.

FINANCIAL EVALUATIONS

This section is designed to give the reader a review of the techniques of financial evaluation and their application. It is not designed to train the reader in all of the characteristics or uses of these techniques. There are other more technical books written for that purpose.

For the construction firm the evaluation of alternative equipment purchased is a must. In technical terms, every firm faces the problem of capital rationing. To choose between alternatives, the firm's managers must estimate the cash outflows and inflows from each investment, calculate the net cash flow, and apply the financial evaluation techniques (described below) to the data. If an equipment item is to be purchased regardless, that is, for other reasons, the financial impact should be calculated and known by the managers who are to make the decision. If the effect is a 5% dilution of income versus a 35% dilution of income, in most construction and other firms the decision would not be the same.

These techniques can also be applied to such decisions as an investment in a computer system versus the hiring of more es-

---

[8]Charles T. Horngren, *Cost Accounting: A Managerial Emphasis* (Englewood Cliffs, New Jersey: Prentice-Hall, 1977), pp. 225-236.
[9]*Ibid.*, pp. 338-345.

timators. For this decision, the first step is to estimate the net cash flow from each alternative. The alternative yearly cash flows are subject to discounting techniques to change them to present values. Then the measurements for each alternative are calculated (from discounted values). From this information the alternative desired is selected (also the reason for each alternative is part of the input to the decision).

DISCOUNTED CASH FLOW

The purpose of the discounting of cash flows is to consider the effect of the passage of time on the value of money, that is, a dollar. This technique can be used to account for the effects of inflation, but that is not the prime purpose of its design. It was designed to recognize interest or a return that can be earned on an investment. Ignoring inflation or other types of returns, if one has a dollar today it is worth more than a dollar received next year because it (the dollar) can be put in the bank to earn interest at the going rate. Because of interest, a dollar received today is always worth more than a dollar received at some time in the future.

One of the discounting techniques (measurements—statistics) that is widely used is the discounted cash flow return. This measurement is expressed as a percentage return, such as 10%. It is calculated by randomly discounting at assumed rates the cumulative cash inflows and outflows of an alternative. The point where the cumulative cash inflows and outflows discounted are equal to zero is the DCF return for that alternative. The percentage return for each alternative can be compared when a choice is necessary. This measurement assumes that the cash flow received during the life of the project (alternative) can be invested at a return at least equal to that of the project itself.

The net present value is another measurement widely used for financial evaluations. Each alternative cash flow is discounted at the same discount rate (which may be an interest rate) over the life of the project. The net cash flow for each year is added (after discounting) and the result is the net present value of that alternative. The net present values for each alternative can be compared for choosing the project for investment. A net present value index can be calculated by dividing the investment required into the net present value of the project. This index can help to overcome the weaknesses in the use of net present value if the alternatives are not equal in the amount of investment required or in the life of the project. The discount rate is often based on the firm's cost of capital. A method for calculating the cost of capital is discussed below.

Another investment measurement is the payback period. The payback period is the year in the future (or period in the future) when the investment will be returned to the firm. The payback can be calculated raw or discounted. The recommended method is the discounted payback period because it adjusts for the time value of money. This measure is particularly useful when a firm has a shortage of cash. The alternative is chosen that pays the investment back the quickest.

THE COST OF CAPITAL AND INDEXING

The cost of capital to the firm is the basis for determining the discount rate to be applied in financial analysis (calculating net present value). One method for calculating the cost of capital is to use the latest financial statements. It can be calculated before or after income taxes from the right side of the statement of financial position (balance sheet), that is, liabilities and owner's equity. It is the cost of all funds to the construction firm. That includes the cost of open payables from suppliers, short term loans, long term loans and mortgages, and capital supplied by owners.

Once the cost of all capital to the firm has been calculated, it becomes the basis for the discount or the return expected from investments. It is the cutoff rate. The return expected will not necessarily be identical to the cost of capital, but the cost of capital is a rate below which the firm will not go. It does not want to earn less than the capital costs and usually the rate is something above the estimated cost of capital to the firm.

Many firms use a project or investment ranking index to aid managers when choosing between alternatives. The index is calculated by dividing the firm's objective (the cost of capital or cutoff or discount rate of return) by the projects discounted rate of return. By this technique projects can be ranked numerically. A typical report for managers in choosing the investment alternative is:

Description (words) and Project #	Ranking Index	NPV	DCF Return	Discounted Payback Period
New asphalt spreader —will spread asphalt one and one-third faster than equipment it is replacing—#81100001	1.34	$56,000	29%	$2\frac{1}{2}$ periods

Each project would be listed on the report by its ranking index: the highest ranking index first and then in descending order.[10]

AUDITS OF FINANCIAL EVALUATIONS

If financial evaluations are used by a firm, maximum benefit is not possible unless the evaluations are audited. The audit measures the accuracy of the evaluation procedure. If, for each evaluation, a manager is assigned responsibility for that evaluation, the audit becomes another input into the evaluation of performance. The project objective is included in the managers Management by Objectives review and the audit is verification of performance. If a manager's projects (evaluations) are continuously inaccurate, corrective action can be taken.

Audits of financial evaluations can be continuous and/or final. Continuous audits are usually planned to start after a specified period into the projects life. For example, if the project life is three years, the first audit may be made after the first year of the project's life. Another audit would be made at the end of the second year and then a final audit at the end of the project's life. The continuous audit results are compared to that portion of the original estimates which applies to that stage of the project's life. Some firms prepare final audits only after or at the end of a project's life. The continuous audits are the most useful to managers because no decision is completely irrevocable.

SUMMARY

This chapter includes a discussion of financial ratios and financial evaluations. The financial ratios discussed are the working capital ratio (sometimes called the current ratio), the quick ratio, the return on stockholder's investment, the asset turn, the margin on sales, and the return on capital employed by the firm. Included in the discussion is a comment on the need for and use of ratios internally by the firm's managers.

The calculation and use of the Z score are described. The description includes the calculation of each component of the Z score and the interpretation of the total score. An example illustrates the application of the Z score to a large firm.

The review of financial evaluations emphasizes the need to

[10] Erich A. Helfert, *Techniques of Financial Analysis*, (Homewood, Illinois: Richard D. Irwin, 1972), pp. 116-127.

classify expenses into fixed and variable categories. These classifications are necessary if the firm is to use contribution bidding. Also presented is the application of these techniques to the choice of equipment and the evaluation of purchases of assets versus the incurrence of an expense.

Discounted cash flow return, net present value, the discounted and raw payback, the cost of capital to the firm, and the calculation of project ranking indices are described in this chapter. Chapter 12 concludes with a discussion of the need for and the uses of audits of financial evaluations.

QUESTIONS

1. Describe and compute the working capital ratio.
2. Who would use the working capital ratio?
3. Calculate the quick ratio (show the formula).
4. Of what use is the quick ratio?
5. What is the meaning of the return on stockholder's investments?
6. How is the return on stockholder's investment calculated?
7. What are the meaning and uses of asset turn?
8. How is asset turn calculated?
9. What is the formula for calculating margin on sales?
10. How should the margin on sales be interpreted and used?
11. Describe and discuss return on capital employed.
12. What is the formula for the calculation of return on capital employed?
13. Discuss the internal use of financial ratios.
14. What are financial evaluations?
15. Why is it necessary to separate expenses into fixed and variable classifications?
16. How are financial evaluations useful for choosing between capital purchases and the expenditure of funds?
17. What is discounted cash flow?
18. Describe the discounted cash flow return.
19. What is net present value?
20. Discuss the raw and discounted payback period.
21. Describe the cost of capital for a construction firm.
22. What are project ranking indices?

23. Describe a final audit of a financial evaluation.

24. What are continuous audits of financial evaluations?

25. How are audits of financial evaluations used?

REFERENCES

Anthony, Robert N., and James S. Reece, *Management Accounting* (Homewood, Illinois: Richard D. Irwin, 1975).

"Bankruptcies—Foresee, Forestall, But Don't Fall Short," *The CPA Journal*, Vol. 49, No. 11, November 1979.

Barfield, J., "Extending Credit in the Construction Industry," *Credit and Financial Management*, Vol. 80, November 1978.

"California Supreme Court Upholds Mechanics Liens," *Air Conditioning and Refrigeration News*, Vol. 139, October 25, 1976.

Chorba, George J., *Accounting for Managers* (New York: American Management Association Extension Institute, 1978).

Clark, R., "Owners of Accounts Receivable of Contractors Analyzed," *National Underwriters* (*Fire Edition*), Vol. 23, July 4, 1969.

Coombs, William E., and William J. Palmer, *Construction Accounting and Financial Management* (New York: McGraw-Hill Book Company, 1977).

Dearden, John, *Cost Accounting and Financial Control Systems* (Reading, Massachusetts: Addison-Wesley Publishing Co., 1973).

Garrison, Ray H., *Managerial Accounting, Concepts for Planning, Control, Decision Making* (Dallas, Texas: Business Publications, 1979).

Gordon, Myron J., and Gordon Shillinglaw, *Accounting: A Management Approach* (Homewood, Illinois: Richard D. Irwin, 1974).

Helfert, Erich A., *Techniques of Financial Analysis* (Homewood, Illinois: Richard D. Irwin, 1972).

Horngren, Charles T., *Cost Accounting: A Managerial Emphasis*, Englewood Cliffs, New Jersey: Prentice-Hall, 1977).

Kieso, Donald E., and Jerry J. Weygandt, *Intermediate Accounting* (New York: John Wiley & Sons, 1977).

Kieso, Donald E., and Jerry J. Weygandt, *Intermediate Accounting* (New York: John Wiley & Sons, 1980).

Meigs, Walter B., A. N. Mosich, and E. John Larson, *Modern Advanced Accounting* (New York: McGraw-Hill Book Company, 1979).

Montgomery, A. Thompson, *Managerial Accounting Information* (Menlo Park, California: Addison-Wesley Publishing Company, 1979).

Mott, Charles H., *Accounting Reports for Management* (Englewood Cliffs, New Jersey: Prentice-Hall, 1979).

Murdick, Robert G., Thomas C. Fuller, Joel E. Ross, and Frank J. Winnermark, Accounting Information Systems (Englewood Cliffs, New Jersey: Prentice-Hall, 1978).

Peterson, J., "Bonding of Contractors—A Surety's Analysis," *The Journal of Commercial Bank Lending*, Vol. 58, July 1976.

"Professional Notes," *The Journal of Accountancy*, December 1979.

Pyle, William W., John Arch White, and Kermit D. Larson, *Fundamental Accounting Principles* (Homewood, Illinois: Richard D. Irwin, 1978).

Rayburn, L. Gayle, *Principles of Cost Accounting with Managerial Applications* (Homewood, Illinois: Richard D. Irwin, 1979).

Welsch, Glenn A., Charles T. Zlatkovich, and Walter T. Harrison, Jr., *Intermediate Accounting* (Homewood, Illinois: Richard D. Irwin, 1979).

Weston, J. Fred, and Eugene F. Brigham, *Managerial Finance* (Hinsdale, Illinois: The Dryden Press, 1978).

CHAPTER 13

SUMMARY AND CONCLUSIONS

This book is designed to train accountants and others who are not familiar with the peculiarities of the construction industry. It is useful for those who have commerce with construction firms as well as those employed in the industry. Emphasis is placed on areas that are different from other industries and require unique exposure or experience.

THE APPLICABILITY OF GENERALLY ACCEPTED ACCOUNTING PRINCIPLES

The rules of the accounting profession require that construction firms conform to generally accepted accounting principles in the preparation and reporting of accounting and financial information. The classifications and structure of its external reports must meet these standards. However, generally accepted accounting principles allow industry exceptions. The construction industry is one of these acceptable exceptions.

The exceptions are principally in the area of revenue recognition and the related receivables. Depending on the method of revenue recognition selected by the construction firm, exchange may not necessarily have occurred. This exception requires extra diligence on the part of the preparer and reviewer of financial information and reports. The reader must also recall that generally accepted accounting principles apply only to financial information prepared for external use and not to the preparation of income tax returns. In addition, the firm can choose one method for external reporting and another method for income tax preparation.

The result is financial reports and accounting systems that on the whole conform to the requirements of the accounting profession but have areas of variation. The preparer, reviewer, and user of the information must be aware of the exceptions and their effect.

FINANCIAL CONTROL IN THE CONSTRUCTION INDUSTRY

Financial controls in the construction industry are more difficult to design, install, and enforce than in other industries. The need, however, is just as critical, if not more so. Contributing factors are the many and remote locations in which most construction firms operate. The duplication at all locations of the usual financial controls would be too costly. The need to meet deadlines and the causes of possible delays are multiplied by the duplicity of locations and, sometimes, their remoteness.

As a result, heavy reliance is placed on field personnel and field supervision. This reliance, in effect, puts the burden of job completion and of financial responsibility on field employees. Because of the broad responsibilities of field personnel in the preparation of time and job cards, purchase orders, requisitions, and receiving reports, they are subject to tremendous temptation.

Often one part of their responsibility is performed less well than other parts and, because of the training and reporting relationships of field personnel, it is usually the financial responsibilities and controls that are overlooked. This causes a further breakdown in controls and additional temptations. At best, only controls after the fact and over very large expenditures are achieved.

When the effect of the weather on construction performance is added to the usual strains in internal controls, the result is a complete breakdown of controls, particularly at the end of the construction season. To offset the pressure on and the violation of internal controls, the construction firm can put more resources into their maintenance. The objective is to fund financial controls until the incremental costs of the controls equal the possible or expected loss from lack of controls. In most cases, the reality is a movement around equilibrium rather than the attainment of it.

CASH FLOW IN THE CONSTRUCTION INDUSTRY

Cash flow is usually less smooth in the construction industry than in manufacturing. The construction cycle is often longer than the manufacturing cycle and there are no product returns. Thus the final inspection process becomes extremely important and the acceptance of the project by the customer is paramount.

Because of this peculiarity (the construction firm improves real property), a tradition of partial payments throughout the construction process and retention of funds until completion has developed. Partial payments are made at certain percentages of completion of phases of the construction contract as authorized by the contract. Withheld from these payments is a percentage for retention which is paid at completion of the phase, project, or total contract. The result is a reduction in the construction firm's cash flow.

THE NEED FOR COMMUNICATION BETWEEN FINANCE AND FIELD PERSONNEL

Communication is required in all industries. Because of the reliance on field personnel for most of their information, however, the financial function in the construction industry is particularly vulnerable to a failure to communicate. Field personnel provide information on the receipt of construction material and sometimes purchase it. They provide information on the hours worked by construction crews and on the projects or jobs to which those hours apply.

Field personnel, particularly field supervision, receipt the waybills of haulers. Field personnel are responsible for the evaluation, performance, and certification of subcontractor invoices. Field personnel report the percentage completion of phases of construction contracts for the preparation and submission of partial invoices (billings). They are also instrumental in the acceptance of a completed project by the customer. Field personnel are responsible for errors, omissions, and missed completion dates, all of which affect the construction firm's cash flow.

The effect of the actions and decisions of field personnel on the construction firm's cash flow makes continuous and excellent communications between financial personnel and field personnel a must. Often field personnel are unaware of their role and effect on the firm's cash flow. Temporary arrangements for borrowing cash or delaying repayment of a debt are possible if the cash shortage is predictable. Communication from field personnel and feedback reporting by finance are necessary to predict cash flow. Without adequate cash flow the firm will fail or at best perform poorly.

All of these peculiarities of the construction industry are illustrated in detail in this book.

QUESTIONS

1. Define an operational asset.
2. At what value are operational assets recorded?
3. How is the purchase of an asset evaluated?
4. What techniques can be used to provide the necessary physical control of operational assets?
5. What is the journal entry for recording depreciation?
6. Describe the units of production method of recording depreciation.
7. Describe the straight line method of depreciation.
8. Illustrate and describe the sum of the years digits method of recording depreciation.
9. Illustrate and describe the double the straight line rate on the declining balance method of depreciation.
10. Discuss and describe the composite methods of recording depreciation.
11. How are the intervals for recording depreciation chosen?
12. What journal entries and methods are used to identify depreciation charges with projects?
13. Discuss insurance in the construction industry.
14. What journal entries are required to record insurance costs?
15. What are real and personal property taxes and how are they recorded?
16. Prepare the journal entry to record real and personal property taxes.
17. What is the journal entry to record federal, state, and subdivision income taxes?
18. Describe inventory taxes.
19. How are inventory taxes recorded?
20. Discuss and describe sales and use taxes.
21. What problems can sales and use taxes cause if not properly covered in the construction contract?
22. Prepare the journal entry to record the liability for sales and use taxes.

INDEX